Applied Time Series Analysis for the Social Sciences

Applied Time Series Analysis for the Social Sciences

Specification, Estimation, and Inference

Regina M. Baker

WILEY

This edition first published 2026

Registered Offices
John Wiley & Sons, Inc., 111 River Street, Hoboken, NJ 07030, USA
John Wiley & Sons Ltd. New Era House, 8 Oldlands Way, Bognor Regis, West Sussex, PO22 9NQ, UK

For details of our global editorial offices, customer services, and more information about Wiley products visit us at www.wiley.com.

The manufacturer's authorized representative according to the EU General Product Safety Regulation is Wiley-VCH GmbH, Boschstr. 12, 69469 Weinheim, Germany,
e-mail: Product_Safety@wiley.com.

Wiley also publishes its books in a variety of electronic formats and by print-on-demand. Some content that appears in standard print versions of this book may not be available in other formats.

Library of Congress Cataloging-in-Publication Data Applied for:
Hardback ISBN: 9780470749937

Cover Design: Wiley
Cover Images: abstract background © enjoynz/Getty Images,
Graph © G. E. P. and Jenkins, G. M. (2015), canonical LSE school paper © James E. H. Davidson, David F. Hendry, Frank Srba and Stephen Yeo (1978), VAR model © Christopher A. Sims and Tao Zha (1999)/full credits inside

Set in 9.5/12.5pt STIXTwoText by Straive, Chennai, India
Printed and bound by CPI Group (UK) Ltd, Croydon, CR0 4YY

C9780470749937_041125

Contents

Acknowledgments

First and foremost, I'm deeply grateful to my former students. Many were social scientists without strong math backgrounds, and their intelligence and inquisitiveness were a powerful drive behind my attempt to find clearer ways of explaining the concepts behind time series analysis without burying them in advanced mathematics.

The majority of them were students in the time series courses I taught for the Interuniversity Consortium for Political and Social Research (ICPSR) Summer Program. It was an honor to be part of such a distinguished faculty and a vaunted training program that drew students from all over the world. Thank you, Hank Heitowit, for providing that opportunity.

Before I was promoted to lecturer for the course, I was a teaching assistant for three fascinatingly different but stellar time series professors at ICPSR. Greg Markus made the tricky fundamentals of dynamic equations seem intuitive. B. Dan Wood specialized in ARIMA models at the time and made sure that his students understood every fine detail of sound analysis and interpretation. Jim Granato was an enthusiastic supporter of the London School of Economics (LSE) approach and responsible for introducing many of us who thought we already knew time series to these newer methodologies and to Vector Autoregression.

Professor Markus was generous enough to read several draft chapters long after he hoped to be happily retired. John Elder kindly answered my cold call and provided helpful, thorough advice far beyond what I could reasonably ask, which was merely a check to make sure I wasn't mischaracterizing his work.

Don Eckford was a wonderful proof-reader, replicated my examples in R, and provided the code for the website. His skills are matched only by his patience and good humor.

Finally, it was mortifying to be the delinquent writer always asking for extensions, and I marvel at the patience and kindness of my editors at Wiley – in particular, Kimberly Monroe-Hill and Paul Sayer.

All errors and shortcomings are my own.

About the Companion Website

This book is accompanied by a companion website:

www.wiley.com/go/baker

The website includes:

- Appendices
- Data
- Sample code for chapters 3, 5, 6, 7, 8, and 12
- Exercises for chapters 1, 2, 3, 4, 5, and 10

1

Introduction

1.1 Why Time Series and Why This Book?

Consider the following problems:

1) In testimony to US congressional committees in September of 2007, General David Petraeus noted that "the overall number of security incidents has declined in 8 of the past 12 weeks."[1] A graph of US troop fatalities in Iraq beginning in February of 2007, when the surge began, is consistent with General Petraeus's claim and appears at first glance to provide dramatic support for the belief that the surge was effective; see the graph on the left-hand side of Figure 1.1. The graph on the right-hand side of the figure, though, raises questions about whether the declining number of deaths should be attributed to the surge. While it is true that there was a dramatic decrease in the number of US troop fatalities immediately after the beginning of the surge, there were equally dramatic decreases– and increases– in the number of troop fatalities at other times. Indeed, the series' volatility is apparent, raising the possibility that the reduction attributed to the surge might just as reasonably be attributed to chance variation. In this case, the use of time series techniques is motivated by an important research design issue.
2) Studies of the relationship between politicians' electoral success and the economic well-being of their constituents are a staple in the voting behavior literature. Researchers attempting to extend these analyses by election results over time may find that the residuals from ordinary least squares

1 General David H. Petraeus, "Report to Congress on the Situation in Iraq." 10–11 September 2007.

Applied Time Series Analysis for the Social Sciences: Specification, Estimation, and Inference, First Edition. Regina M. Baker.

Companion website: www.wiley.com/go/baker

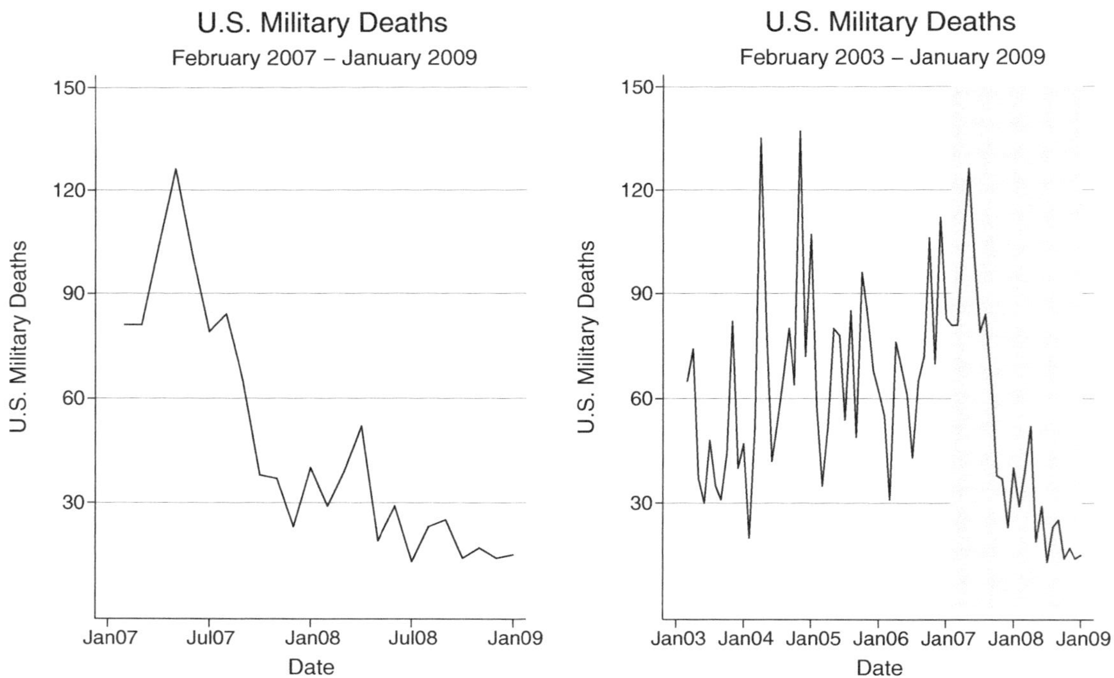

Figure 1.1 US troop fatalities in Iraq, 2003–2009

(OLS) regression are autocorrelated. While introductory regression texts are likely to advise the use of generalized least squares (GLS) in this case, doing so ignores the possibility that interesting omitted variables are causing the autocorrelation. Reflexively applying GLS not only leaves that possibility unexplored, but will lead to biased estimates of the coefficients of interest if the omitted variables are also associated with the included variables.

3) Finally, researchers often find themselves with time series data relevant to their substantive interests, have some sense that they should use time series methods, but do not have the necessary training.

These common motivations for using time series techniques differ critically from the motivation assumed in most other time series presentations: the researchers did not begin with an attachment either to a particular time series technique, or to a theoretically-derived equation representing the relationship between variables measured at different time periods. Instead, methods such as ARIMA[2] modeling, generalized least squares regression, vector autoregression and autoregressive distributed lag models (including cointegration and error correction specifications) are presented separately, providing little information about the connections between them and the reasons for choosing one over another. Additionally, many econometricians begin with the assumption that their theory is correct – that they have included all relevant variables and controls – and seek only to estimate the model's parameters.

Developments in time series analysis over the last few decades suggest that this process may be more of a hindrance to good time series work than a help. In the same way that researchers working with cross-sectional data risk misinterpreting biased estimates if they do not consider possible confounding variables (omitted variables that are associated with both the dependent variable and one or more included explanatory variables), researchers working with time series data risk biased estimates if they do not utilize modeling techniques that give appropriate attention to delayed effects and complicated dynamic relationships. After some introductory remarks about time series data, a simple illustration will be provided in the next section.

2 ARIMA is an acronym for "Autoregressive Integrated Moving Average." These models will be presented in detail in Chapters 4–7.

1.2 Time Series: Preliminaries

1.2.1 What is a Time Series?

The critical feature of time series data is that observations have a logical order. For example, a subset of the Penn World Tables is presented in Table 1.1. Population, exports and imports are each a time series, and the accumulated data are logically ordered by the years for which they are reported. Ideally, time series observations are equally spaced – common examples include yearly data, monthly data, quarterly data, and daily data.

Notice that these data refer to a single unit (country, in this case) – the United States. Data collected over time for more than one country, as in Table 1.2, are referred to as "Pooled Cross-section Time Series" data, and require attention to issues beyond those presented in this book.

Table 1 .1 An example of time series data.

Country	Year	Population	Exports (in millions)	Imports (in millions)
United States	1950	152 271 000	55 391	65 172
United States	1951	154 878 000	67 491	67 688
United States	1952	157 553 000	63 319	73 601
United States	1953	160 184 000	61 024	80 521
United States	1954	163 026 000	64 049	78 005
⋮	⋮	⋮	⋮	⋮
United States	2007	301 279 593	153 409	218 379

Exports and imports are shown in current international prices.
(Source: Heston et al., (2009)).

Table 1.2 An example of pooled cross-section time series data.

Country	Year	Population Population	Exports (in millions)	Imports (in millions)
Canada	1950	14 011 422	27 902	23 190
Canada	1951	14 330 675	30 497	26 708
Canada	1952	14 785 584	33 255	27 668

(Continued)

Table 1.2 (*Continued*)

Country	Year	Population Population	Exports (in millions)	Imports (in millions)
Canada	1953	15 183 375	32 930	30 067
Canada	1954	15 636 245	31 795	28 467
⋮	⋮	⋮	⋮	⋮
Canada	2007	32 935 961	52 895	51 945
Mexico	1950	28 485 180	43 307	74 556
Mexico	1951	29 296 235	40 583	93 704
Mexico	1952	30 144 317	42 059	91 260
Mexico	1953	31 031 279	48 927	87 186
Mexico	1954	31 959 113	51197	91 260
⋮	⋮	⋮	⋮	⋮
Mexico	2007	108 700 891	294 000	319 500
United States	1950	152 271 000	55 391	65 172
United States	1951	154 878 000	67 491	67 688
United States	1952	157 553 000	63 319	73 601
United States	1953	160 184 000	61 024	80 521
United States	1954	163 026 000	64 049	78 005
⋮	⋮	⋮	⋮	⋮
United States	2007	301 279 593	153 409	218 379

Exports and imports are shown in current international prices.
(Source: Heston et al., (2009)).

1.2.2 Why Special Methods?

Many students leave introductory regression courses with the impression that proper estimation of time series models requires only attention to autocorrelation and that this is readily accomplished through the use of generalized least squares or Newey–West estimators. Autocorrelation, though, is often a symptom of a larger misspecification issue – a failure to properly model the dynamic process generating the data. Special techniques allow researchers to use the evidence the autocorrelation provides to better account for omitted variables, to allow for the possibility of delayed and/or long-lasting impacts of explanatory variables, and to recognize the mathematical equivalence of models that appear to suggest very different causal stories.

1.2.2.1 Autocorrelation and Omitted Variables

First, consider a simple example using simulated data, generated as follows

$$C_t = 5 + 0.3C_{t-1} + 0.9Y_t + u_t,$$

where u_t is a stochastic error term, $u_t \sim N(0, 1)$.

The results of two OLS regressions are reported in Table 1.3. In the first, the model is correctly specified – i.e., the equation is estimated as

$$C_t = \beta_0 + \beta_1 C_{t-1} + \beta_2 Y_t + u_{1t}. \tag{1.1}$$

In the second, though, C_{t-1} is omitted from the specification:

$$C_t = \beta_0 + \beta_2 Y_t + u_{2t}. \tag{1.2}$$

Notice the difference in Durbin–Watson (DW) statistics. The DW statistic for the correctly specified model is very close to 2, indicating that there is little or no autocorrelation in the error term. The DW statistic for the model in which C_{t-1} is omitted, however, is 1.32, suggesting autocorrelation of approximately 0.34.[3]

This case is relatively benign. While omitting C_{t-1} from the equation affected standard errors and the estimate for the constant term,[4] the coefficient estimated for Y_t in the misspecified model is estimated as 0.89, here, very close to the true value of 0.90.

Table 1.3 Results for OLS estimations of simulated data.

Equation	β_0	β_1	β_2	SE	DW
1.1	5.02	0.30	0.90	1.06	1.97
	(0.14)	(0.01)	(0.01)		
1.2	7.65		0.89	3.42	1.32
	(0.34)		(0.03)		

Standard errors are reported in parentheses.
SE is the standard error of the regression.
DW is the Durbin–Watson statistic.

3 The Durbin–Watson statistic is discussed in detail in Chapter 8. Briefly, it ranges between 0 and 4, with 0 indicating maximum positive autocorrelation, 1.0; 2 indicating autocorrelation of 0; and 4 indicating maximum negative autocorrelation, –1.0

4 The change in the estimate for the constant term will be discussed in the Exercises to Chapter 2.

In real data, though, omitted variables are often correlated with included variables, and so biased coefficients must also be a concern. Consider a second example using monthly data for consumption (C_t) and income (Y_t) in the United States, January 1959–June 2012.[5] In this case, of course, we do not know the true model, but for the sake of illustration, estimate the same models as before, Equations (1.1) and (1.2). The results are presented in Table 1.4.

Here also, it appears that autocorrelation is a much more serious problem when C_{t-1} is omitted from the model than when it is included, and again, estimates of the constant term differ substantially. Now, though, estimates of β_2, the coefficient on Y_t, are also very different: 0.93 in the model omitting C_{t-1}, but only 0.02 in the model including C_{t-1}.

Researchers accustomed to working with cross-sectional data might infer that the smaller estimate is correct, and the intuition that C_{t-1} is an important control variable is sound. However, this need not be due to a direct impact that past consumption has on current consumption. Instead, C_{t-1} may be a valuable proxy for any number of unmeasured "causes" of current consumption.

Suppose, for example, that the true model is

$$C_t = \beta_0^* + \beta_2^* Y_t + \beta_3 X_t + u_{2t},$$

where X_t is consumers' exuberance – some unmeasured psychological propensity to buy things. Lagging this equation once,

$$C_{t-1} = \beta_0^* + \beta_2^* Y_{t-1} + \beta_3 X_{t-1} + u_{2t-1}.$$

Table 1.4 Results for OLS estimations of real data.

Equation	β_0	β_1	β_2	SE	DW
1.1	3.83	0.97	0.02	23.4	1.95
	(1.67)	(0.01)	(0.01)		
1.2	–69.5		0.93	81.0	0.26
	(4.90)		(0.0009)		

Standard errors are reported in parentheses.
SE is the standard error of the regression.
DW is the Durbin–Watson statistic.

5 Source: US Department of Commerce, Bureau of Economic Analysis. Downloaded from www.research.stlouisfed.org 16 August, 2012. Both the consumption series and the income series are seasonally adjusted annual rates. Full documentation for these variables, PCE and DSPI, respectively, are provided at www.bea.gov/national/pdf/nipaguide.pdf.

That is, C_{t-1} contains information about consumers' exuberance in the previous period even though we are unable to measure this exuberance directly! To the extent that exuberance is autocorrelated – that is, to the extent that the last period's exuberance is a good predictor of the current period's exuberance – including C_{t-1} in the estimating equation provides some measure of control for a completely unmeasured omitted variable.

Thinking of autocorrelation as merely a nuisance to be swatted away with a statistical fix, then, can be a serious mistake. Clearly, as with cross-sectional data, correctly specifying the model by including all relevant variables is always the best strategy. In many cases, though, analysts either do not know they are omitting important variables or cannot measure them. Carefully considering the reason for autocorrelation and choosing the appropriate time series technique is a critical step toward minimizing omitted variables bias.

1.2.2.2 Delayed Effects

Additionally, independent variables may exert effects over several time periods, or exert effects only after a delay. In these cases, lags of the independent variables should be included in addition to, or instead of, the contemporaneous version of the variable. For the sake of illustration, consider a very simple model exploring the relationship between voters' approval of presidential performance, and changes in real disposable income. Researchers accustomed to working with cross-sectional data are likely to begin with a model like the following:

$$\text{PresApp}_t = \alpha + \beta \Delta \text{Income}_t + U_t \tag{1.3}$$

where α and β are parameters to be estimated and U_t is a random error term.

In the case of time series analysis, though, it is important to remember what is implied by the inclusion of only the contemporaneous income variable: the researcher is assuming that voters' approval is determined *de novo* each period – i.e., that any memory of changes in income during previous periods is irrelevant to current assessments of presidential performance. Taking past performance into account would entail adding lagged income variables to the equation:

$$\text{PresApp}_t = \alpha + \beta_0 \Delta \text{Income}_t + \beta_1 \Delta \text{Income}_{t-1} + \beta_2 \Delta \text{Income}_{t-2} + U_t^*. \tag{1.4}$$

Time series techniques can aid researchers in evaluating the timing of effects and therefore the number of lags to include.

That said, lagged versions of variables are typically highly correlated – changes in income in one period, for example, tend to be correlated with changes in income in the previous period, which leads to a third important reason for having a solid understanding of dynamic models before doing applied work in time series.

1.2.2.3 Causal Representation

In particular, problems such as the near-collinearity issue described in the previous section may be addressed by estimating a model which is mathematically equivalent, but appears at first glance to suggest a different causal mechanism. Indeed, a common mistake made by researchers accustomed to working with cross-sectional data concerns the interpretation of time series models – in particular, the tendency to treat specifications as literal representations of causal hypotheses, and to treat estimated coefficients as if they represent the magnitudes of hypothesized effects. In beginning regression courses, after all, researchers are typically taught to think of the right-hand-side variables in their models as hypothesized causes of the left-hand-side (dependent) variable. Good researchers remember that correlation does not prove causation, and are careful to temper any causal inferences with appropriate consideration of issues such as selection bias, reverse causality, and confounding variables. With the appropriate caveats in mind, though, estimated coefficients are reasonably interpreted as "marginal effects."

The interpretation of time series models is not nearly so straightforward. Consider, for example, the following two models:[6]

$$\text{PresApp}_t = \beta_0(1-\beta_1) + \beta_1\text{PresApp}_{t-1} + \beta_2\Delta\text{Income}_t + \epsilon_t - \beta_1\epsilon_{t-1}, \tag{1.5}$$

and

$$\begin{aligned}\text{PresApp}_t = \beta_0 + \beta_2\Delta\text{Income}_t + \beta_1\beta_2\Delta\text{Income}_{t-1} \\ + \beta_1^2\beta_2\Delta\text{Income}_{t-2} + \beta_1^3\beta_2\Delta\text{Income}_{t-3}\ldots + \epsilon_t\end{aligned}$$

or, in more compact form,

$$\text{PresApp}_t = \beta_0 + \sum_{k=0}^{\infty}\beta_1^k\beta_2\Delta\text{Income}_{t-k} + \epsilon_t. \tag{1.6}$$

An analyst unaccustomed to working with time series data might believe that Equations (1.5) and (1.6) represent very different causal stories because, for example, PresApp_{t-1} appears in Equation (1.5) but not in Equation (1.6). They are, however, mathematically equivalent statements, and recognizing such equivalences is critical to good time series analysis for at least three reasons. First, estimating the two models will not allow the researcher to discern between the different causal stories. Second, Equation (1.5) implies

$$\text{PresApp}_{t-1} = \beta_0(1-\beta_1) + \beta_1\text{PresApp}_{t-2} + \beta_2\Delta\text{Income}_t + \epsilon_{t-1} - \beta_1\epsilon_{t-2}.$$

---

6 In the following equations, ϵ_t is assumed to be normally and independently distributed with mean 0 and variance σ^2.

But if $\Delta\text{Income}_{t-1}$ exerts an effect on PresApp_{t-1}, and PresApp_{t-1} exerts an effect on PresApp_t, then $\Delta\text{Income}_{t-1}$ has an indirect effect on PresApp_t. It would be a mistake, then, to believe that the effect of "ΔIncome" on PresApp is estimated simply as β_2. In Chapter 9, you will learn that a more reasonable estimate of this effect is $\beta_2/(1-\beta_1)$. Finally, the mathematical equivalence of models which appear to be very different can be exploited for estimation purposes. Because $\Delta\text{Income}_{t-1}$, $\Delta\text{Income}_{t-2}$, etc. are likely to be highly correlated, obtaining precise estimates of the coefficients in Equation (1.6) may be prohibitively difficult. Estimating the mathematically equivalent (1.5) may solve the problem.

In sum, specifying models for time series data involves three considerations not present when specifying models for cross-sectional data. First, autocorrelation may indicate a problem with omitted variables that can be addressed through the use of a good proxy. Second, researchers must consider the timing of effects – whether or not there is a delay before an effect is observed, whether the effect persists for more than one period, etc. Finally, the mathematical equivalence of models which appear to represent very different causal stories must be taken into account when trying to assess the usefulness of competing theories, but may also be exploited to obtain estimates of models that could not be precisely estimated directly.

1.3 Time Series Approaches: Some History, and Outline of the Book

This chapter has approached time series from the perspective of researchers trained in regression analysis. Modern time series techniques, though, draw from two traditions: regression approaches and classical time series analysis (alternately referred to as ARIMA or Box–Jenkins analysis). The presentation of topics in this section follows roughly the order in which they became prominent in the econometrics literature.

1.3.1 Regression Approaches and Structural Models

A word about terminology is in order here. In disciplines such as sociology and psychology, the model presented in this section would be referred to as a "simultaneous equations model," while the term "structural model" is reserved for simultaneous equations models with latent variables, alternatively referred to as LISREL models. In the economics literature, however, the term "structural model" is used to refer to simultaneous equations models of economies. This paradigm, advanced in particular by the Cowles Commission of the University of Chicago,[7] was characterized by multiple regression equations reflecting economic theory concerning actors' decisions. Klein's model for the US economy, 1920–1941 is a typical example[8]

$$\text{Cons}_t = \beta_{10} + \beta_{11}\text{Profits}_t + \beta_{12}(W_{\text{pvt}} + W_{\text{gov}})_t + U_{1t} \tag{1.7}$$

$$\text{Inv}_t = \beta_{20} + \beta_{21}\text{Profits}_t + \beta_{23}\text{Profits}_{t-1} + \beta_{24}\text{Cap}_{t-1} + U_{2t} \tag{1.8}$$

$$\begin{aligned} W_{\text{pvt}_t} = \beta_{30} &+ \beta_{31}(\text{Output} + \text{Tax} - W_{\text{gov}})_t \\ &+ \beta_{32}(\text{Output} + \text{Tax} - W_{\text{gov}})_{t-1} + \beta_{33}t + U_{3t} \end{aligned} \tag{1.9}$$

$$\text{Output}_t = \text{Cons}_t + \text{Inv}_t + \text{Gov}_t - Tax_t \tag{1.10}$$

$$\text{Profits}_t = \text{Output}_t - W_{\text{pvt}_t} - W_{\text{Gov}} \tag{1.11}$$

$$\text{Cap}_t = \text{Cap}_{t-1} + \text{Inv}_t \tag{1.12}$$

where

- Cons_t represents consumers' aggregate expenditure in the period t,
- Profits_t and Profits_{t-1} can be thought of as consumers' non-wage income,
- W_{pvt} is consumers' income from wages paid by private firms,
- W_{gov} is consumers' income from wages paid by the government,
- Inv_t represents net investment,
- Cap_{t-1} is the previous year's capital stock,
- Tax represents business taxes,
- the β's are parameters to be estimated, and
- the U_t's are the sum of the omitted variables and stochastic errors.

Typical characteristics of these models include the following:

- Each equation is a fairly literal representation of some theoretical postulate. In this case, the first three represent decisions made by consumers, investors, and employers; the last three are economic accounting identities.
- Causal order is assumed. For example, the private sector wage, W_{pvt_t}, is assumed to be affected by the level of production, Output_t; in Equation (1.9), the private sector wage is the dependent variable and level of production is an independent variable. By assumption, though, the level of production is assumed not to be affected by the level of wages – that is, the private sector wage does not appear as an independent variable in the Equation (1.10).

7 See Hildreth, 1986.
8 See Klein, 1950.

Notice that these are common strategies for specifying models for cross-sectional data. Time series researchers found them lacking, however, for several reasons.

The models are notoriously fragile. It is not uncommon to find that seemingly minor changes to their specifications – for example, dropping and/or adding a single variable – can result in coefficient estimates that are not only quite different in magnitude, but worse, differently signed. Comfort with inferences concerning whether an explanatory variable has a positive effect or a negative effect on a dependent variable seems a minimal requirement for the usefulness of an estimation technique. Moreover, researchers often have strong incentives to add or remove variables in ways that are not consistent with theory.[9] Edward Leamer famously summarized disenchantment with the Cowles Commission approach this way:

> After three decades of churning out estimates, the econometrics club finds itself under critical scrutiny and faces incredulity as never before. Fischer Black writes of "The Trouble with Econometric Models." David Hendry queries "Econometrics: Alchemy or Science?" John W. Pratt and Robert Schlaifer question our understanding of "The Nature and Discovery of Structure." And Christopher Sims suggests blending "Macroeconomics and Reality."
>
> … I too am troubled by the fumes which leak from our computing centers.
>
> (Leamer, 1983:42–43)

Aesthetics were the least of the problems, though. In the early 1970s, researchers dealt a critical blow to the structural modeling approach with papers demonstrating that atheoretical Box–Jenkins techniques yielded more accurate forecasts than the elegantly theory-driven structural models.[10] Forecasting, of course, is important for reasons beyond a sort of predicting-the-future wizardry: a model that is consistent with observations from one time period, but a poor fit for observations from the next, is unlikely to be a good model. The questionable performance of the structural models motivated the development of newer methodologies.[11]

9 The problem of achieving "identification" will be explained in Chapter 11.
10 See, for example, Nelson, 1972 and Cooper, 1972.
11 See Pagan, 1987 and Diebold, 1998 for more detailed summaries.

1.3.2 Classical Time Series: ARIMA (Box–Jenkins) Models

To understand why the superior forecasting performance of ARIMA models presented a crisis, one need only compare the ARIMA and Cowles Commission approaches to modeling economic variables. The most general univariate ARIMA models include only lags of the dependent variable, an error term, and lags of the error term. Rather than forecast future consumption based on an estimation of the system of Equations (1.1)–(1.6), then, one might base the forecast on equations like the following:

$$\text{Cons}_t = \mu + \rho\text{Cons}_{t-1} + e_t \tag{1.13}$$

or

$$\text{Cons}_t = \mu + \rho\text{Cons}_{t-1} + e_t + \theta e_{t-1} \tag{1.14}$$

or

$$\text{Cons}_t = \mu + \rho_1\text{Cons}_{t-1} + \rho_2\text{Cons}_{t-2} + e_t \tag{1.15}$$

or even

$$\text{Cons}_t = \mu + e_t + \theta_1 e_{t-1} + \theta_2 e_{t-2} \tag{1.16}$$

where

$$e_t \sim N(0, \sigma^2) \tag{1.17}$$

and the σ^2, μ, ρ_k and θ_k are parameters to be estimated.

Notice that the estimation of Equations (1.13)–(1.17) requires information only for the consumption series, while Equations (1.7)–(1.12) include information about several other economic variables, such as prices and wages. The superior forecasting performance of the ARIMA model is completely counter-intuitive, since it appears to use much less information than the structural model does.

The superior forecasting performance should also provide some reassurance to those concerned about the atheoretical modeling process. Rather than begin with equations suggested by theory, the selection of models like those represented by Equations (1.13)–(1.17) is based on the pattern of autocorrelations found in the data – in this case, the consumption series – and goodness-of-fit criteria.

These very simple ARIMA models will be introduced in Chapters 4 and 5. While univariate ARIMA models have fallen out of vogue, the insights behind them provide an important foundation for two contemporary approaches that will be studied closely – the LSE/Hendry Approach, and Sims' Vector Autoregression (VAR) models.

Moreover, univariate ARIMA models can be used to establish a baseline against which the effects of an intervention can be measured. Suppose, for example, that I am a social scientist interested in studying the effect of a treaty limiting fishing rights on the population of an endangered species of fish. A simple ARIMA model might be used to model the typical variability in the fish population without requiring careful consideration of predator populations, food sources, and other variables that affect the fish population, but are neither central to my study nor part of my expertise. ARIMA intervention models will be covered in Chapters 6 and 7. The transfer functions discussed in Chapter 6 are a precursor to the impulse response functions central to vector autoregression.

1.3.3 The LSE/Hendry Approach

While contributors to this tradition include Nobel laureates Robert Engle and Clive Granger, both of whom had limited associations with the London School of Economics (LSE), the methodological approach owes much to an tradition developed there under the leadership of David Hendry.[12] Like the ARIMA modelers, researchers in this tradition emphasize the distinction between the data generating process (often abbreviated DGP) and particular realizations, a distinction which should become clearer in Chapter 3.

The standard process involves beginning with a model suggested by theory, making appropriate modifications based on observed time series properties such as (non)stationarity and seasonality, generalizing the model and testing restrictions, and then examining the predictive power of the model to glean insights into appropriate refinements. A brief introduction of the Common Factors (ComFac) critique may give a sense of what is meant by generalizing and testing restrictions. Consider the GLS model frequently recommended for cases in which first-order autocorrelation is observed in the residuals:

$$C_t = \alpha + \beta Y_t + e_t \tag{1.18}$$

where

$$e_t = \rho e_{t-1} + \epsilon_t \tag{1.19}$$

and

$$\epsilon_t \sim N(0, \sigma^2). \tag{1.20}$$

12 Good summaries can be found in Pagan, 1987, Diebold 1990, and Charemza and Deadman 1997.

Using either complicated substitutions or simple shorthand introduced in Chapter 2, it can be shown that Equations (1.18)–(1.19) can be rewritten as the single equation

$$C_t = (1-\rho)\alpha + \rho C_{t-1} + \beta Y_t - \rho\beta Y_{t-1} + \epsilon_t. \tag{1.21}$$

That is, the GLS model is a special case of an autoregressive distributed lag (ADL) model in which there is a particular relationship between the value of the constant term and the coefficients associated with Y_{t-1}, X_t and X_{t-1}. Rather than merely assuming this relationship, general-to-specific modelers advocate estimating the coefficients for the general model

$$C_t = \gamma_1 + \gamma_2 C_{t-1} + \gamma_3 Y_t + \gamma_4 Y_{t-1} + u_t$$

and testing the restriction $\gamma_4 = -\gamma_2\gamma_3$. This methodology is examined more closely in Chapters 8–11. A canonical example of this work is Davidson et al., 1978, which will be discussed in detail in Chapter 11.

1.3.4 Vector Autoregression

The VAR methodology introduced by Christopher Sims takes the general-to-specific modeling recommendation a step further by dropping assumptions regarding the order of causality. To examine the relationship between consumption and income, for example, we would begin by using ordinary least squares to estimate equations taking the following form:

$$C_t = \alpha_1 + \beta_{11} Y_{t-1} + \beta_{12} Y_{t-2} + \ldots \beta_{1k} Y_{t-k} + \gamma_{11} C_{t-1} + \gamma_{12} C_{t-2} + \ldots + \gamma_{1k} C_{t-k} + u_1 \tag{1.22}$$

and

$$Y_t = \alpha_2 + \beta_{21} C_{t-1} + \beta_{22} C_{t-2} + \ldots \beta_{2k} C_{t-k} + \gamma_{21} Y_{t-1} + \gamma_{22} Y_{t-2} + \ldots + \gamma_{2k} Y_{t-k} + u_2, \tag{1.23}$$

where the u_i's are normally distributed with mean 0 and variance σ^2. Tests are used to determine how many lags of each variable to include in each equation. Because of the high correlations between the independent variables, coefficient estimates are imprecise and of little use. Instead, "impulse response functions" (IRF's) are generated showing the effect on one variable, over time, of a change in the other variable. Methods and tests associated with the VAR approach are introduced in Chapters 12.

1.4 Summary

- Many students leave introductory regression courses with the impression that proper estimation of time series models requires only attention to autocorrelation, and that this is readily accomplished with quick statistical fixes.
- Autocorrelation, though, is often a symptom of a misspecified model – e.g., not modeling the dynamic process properly by failing to consider delayed and or lasting effects, or omitting lagged variables that are associated with included variables.
- In the same way that researchers look at scatterplots to decide whether a relationship is nonlinear, time series experts examine time series properties through use of unit root tests, autocorrelation plots and partial autocorrelation plots to make initial judgments about a series' underlying dynamic process and potential omitted variables.
- Unlike the cross-sectional case in which we can only speculate about the effect of omitted confounding variables that cannot be measured directly, time series data often provide good proxies for omitted variables since lagged dependent variables typically include information about the omitted variables.
- Indeed, atheoretical ARIMA models and VAR models have been found to provide better forecasts than models based carefully on theory.

The chapters proceed as follows. Chapter 2 introduces lag operator algebra, a simple tool that makes it easier to see the connections between different model specifications. Chapters 3 and 4 describe time series properties such as nonstationarity and complex autocorrelation structures, and tools for diagnosing them. Chapters 5–7 show how these properties can be used to build atheoretical ARIMA models, and how to use these models as baselines for evaluating the impact of either discrete or continuous explanatory variables. Sections in Chapter 6 explaining transfer functions in ARIMA impact assessment will be useful for understanding the impulse response functions in VAR models. Chapter 7 introduces ARIMA models with continuous explanatory variables.

Chapter 8 evaluates the regression approach to time series in light of the material covered in the earlier chapters. Chapters 9–11 present the more sophisticated regression-based approaches in the spirit of the LSE/Hendry school, including error correction models. Chapter 12 introduces VAR models.

References

Brookings (2008). *Iraq Index: Tracking Variables of Reconstruction & Security in Post-Saddam Iraq, July 17, 2008*. downloaded from http://www.brookings.edu/iraqindex July 19, 2008.

Charemza W. and Deadman D. (1997). *New Directions in Econometric Practice: General to Specific Modelling, Cointegration and Vector Autoregression.* Edward Elgar.

Cooper R.I. (1972). The predictive performance of quarterly econometric models of the United States. In *Econometric Models of Cyclical Behavior* (ed. B.G. Hickman) New York: Columbia University Press.

Davidson J.E.H., Hendry D.F., Srba F., and Yeo S. (1978). Econometric Modelling of the Aggregate Time-Series Relationship Between Consumers' Expenditure and Income in the United Kingdom. *The Economic Journal* 88(352):661–692.

Diebold F.X. (1998). The Past, Present, and Future of Macroeconomic Forecasting. *Journal of Economic Perspectives* 12:175–192.

Heston A., Summers R., and Aten B. (2009). *Penn World Table Version 6.3.* Center for International Comparisons of Production, Income and Prices at the University of Pennsylvania. Downloaded Jan. 22, 2011.

Hildreth C. (1986). *The Cowles Commission in Chicago, 1939–1955.* New York: Springer-Verlag.

Klein L.R. (1950). Economic Fluctuations in the United States, 1921–1941. John Wiley & Sons.

Leamer E.E. (1983). Let's Take the Con Out of Econometrics. *The American Economic Review* 73(1):31–43.

Nelson C.R. (1972). The Prediction Performance of the FRB-MIT-Penn Model of the US Economy. *American Economic Review* 62(5):902–917.

Pagan A.R. (1987). Three econometric methodologies: a critical appraisal. *Journal of Economic Surveys* 1:3–24. Reprinted in *Modeling Economic Series: Readings in Econometric Methodology*. (ed. C.W.J. Granger 1990) Oxford: Oxford University Press.

2

Foundations

Chapter 1 hinted that the interpretation of the estimated coefficients in a time series model with a lagged dependent variable is not straightforward. A prerequisite to understanding time series analysis is recognizing the mathematical equivalence of time series models that seem to suggest very different causal mechanisms. This chapter lays the groundwork.

- The first section focuses on the intuition using simple algebra – the method of successive substitutions.
- In the next set of sections, lag operator algebra is introduced as a tool for making these equivalences more apparent and for simplifying proofs. It is no exaggeration to say that lag operator algebra is as useful for time series analysts as matrix algebra is to those studying systems of linear equations.
- The following section provides an example. The derivation presented in the beginning of the chapter using the method of successive substitutions is derived using lag operator algebra.
- The final section provides a brief aside on difference equations and their solutions. Solving difference equations is a special case of identifying mathematically equivalent equations which appear to suggest very different causal processes.

2.1 Multiple Interpretations: Some Intuition

Consider a simple model of approval of the US President:

$$\text{PresApp}_t = \beta_0 + \beta_1 \text{PresApp}_{t-1} + \beta_2 \text{CDI}_t + \epsilon_t, \tag{2.1}$$

where PresApp_t is the Gallup poll approval rating of the president, CDI is the average change in voters' disposable income, and

Applied Time Series Analysis for the Social Sciences: Specification, Estimation, and Inference, First Edition. Regina M. Baker.

Companion website: www.wiley.com/go/baker

$$\epsilon_t \sim N(0,\sigma^2)$$
$$|\beta_1|<1.$$

Different interpretations are possible because if the model is correctly specified, the lagged dependent variable – i.e., the previous year's Gallup approval rating – contains information about the average change in voters' disposable income going back one more year. This can be represented mathematically by writing the implied equation for the previous time period:

$$\text{PresApp}_{t-1} = \beta_0 + \beta_1\text{PresApp}_{t-2} + \beta_2\text{CDI}_{t-1} + \epsilon_{t-1}.$$

Consequently, some of the "effect" one might attribute to the past period's approval rating might be more appropriately attributed to the past period's CDI. While we think of control variables as permitting us to separate neatly the impact of different "causes" when working with cross-sectional data, such neat separations are typically not possible with time series data. In fact, Equation (2.1) is mathematically equivalent to a model in which the lagged presidential approval variable PresApp_{t-1} is excluded, and only past and present changes in disposable income are included as explanatory variables:

$$\begin{aligned}\text{PresApp}_t = {} & \frac{\beta_0}{1-\beta_1} + \beta_2\text{CDI}_t + \beta_1\beta_2\text{CDI}_{t-1} + \beta_1^2\beta_2\text{CDI}_{t-2} \\ & + \beta_1^3\beta_2\text{CDI}_{t-3} + \beta_1^4\beta_2\text{CDI}_{t-4} + \ldots + v_t\end{aligned}$$

$$\text{PresApp}_t = \frac{\beta_0}{1-\beta_1} + \sum_{k=0}^{\infty}\beta_1^k\beta_2\text{CDI}_{t-k} + v_t, \qquad (2.2)$$

where v_t is a function of the original ϵ_t.

Any concern that we have stumbled upon a mathematical flaw that leads to a contradiction in the substance of our problem should evaporate with closer attention; the ambiguity flows directly from the substance. It is possible that people do not revise their opinions of the President very readily, and so the model written with the lagged dependent variable is a good literal representation of the decision calculus of individuals. It is also possible, though, that people's approval is fully responsive to changes in disposable income, and that the lagged dependent variable PresApp_{t-1} is merely a proxy for the history of changes in disposable income.

To see this using simple algebra, first recognize that if we posit that Equation (2.1) holds for all time periods under consideration, it implies

$$\text{PresApp}_{t-1} = \beta_0 + \beta_1\text{PresApp}_{t-2} + \beta_2\text{CDI}_{t-1} + \epsilon_{t-1} \qquad (2.3)$$

$$\text{PresApp}_{t-2} = \beta_0 + \beta_1\text{PresApp}_{t-3} + \beta_2\text{CDI}_{t-2} + \epsilon_{t-2} \qquad (2.4)$$

$$\text{PresApp}_{t-3} = \beta_0 + \beta_1 \text{PresApp}_{t-4} + \beta_2 \text{CDI}_{t-3} + \epsilon_{t-3} \tag{2.5}$$

$$\vdots$$

$$\text{PresApp}_{t-k} = \beta_0 + \beta_1 \text{PresApp}_{t-k-1} + \beta_2 \text{CDI}_{t-k} + \epsilon_{t-k}, \tag{2.6}$$

where k is any arbitrary number of time periods.

Substituting (2.3) into (2.1),

$$\text{PresApp}_t = \beta_0 + \beta_1(\beta_0 + \beta_1 \text{PresApp}_{t-2} + \beta_2 \text{CDI}_{t-1} + \epsilon_{t-1}) + \beta_2 \text{CDI}_t + \epsilon_t.$$

Rearranging terms,

$$\begin{aligned}\text{PresApp}_t = {} & \beta_0 + \beta_1\beta_0 + \beta_1^2 \text{PresApp}_{t-2} + \beta_2 \text{CDI}_t \\ & + \beta_1\beta_2 \text{CDI}_{t-1} + \varepsilon_t + \beta_1 \varepsilon_{t-1}.\end{aligned}$$

Notice that now that the previous period's presidential approval does not appear as an explanatory variable – instead, presidential approval from two periods back appears. Substituting (2.4) into this new equation and rearranging terms,

$$\begin{aligned}\text{PresApp}_t &= \beta_0 + \beta_1\beta_0 + \beta_1^2(\beta_0 + \beta_1 \text{PresApp}_{t-3} + \beta_2 \text{CDI}_{t-2} + \epsilon_{t-2}) \\ &\quad + \beta_2 \text{CDI}_t + \beta_1\beta_2 \text{CDI}_{t-1} + \epsilon_t + \beta_1 \epsilon_{t-1}) \\ &= \beta_0 + \beta_1\beta_0 + \beta_1^2\beta_0 + \beta_1^3 \text{PresApp}_{t-3} + \beta_2 \text{CDI}_t + \beta_1\beta_2 \text{CDI}_{t-1} \\ &\quad + \beta_2 \text{CDI}_t + \beta_1\beta_2 \text{CDI}_{t-1} + \beta_1^2\beta_2 \text{CDI}_{t-2} + \epsilon_t + \beta_1 \epsilon_{t-1} + \beta_1^2 \epsilon_{t-2}.\end{aligned}$$

Notice what is happening to each of the four components – the constant expression, the expression involving past presidential approval, the expression involving changes in disposable income, and the expression involving random events:

- the expression for the constant term changed from
  - β_0, to
  - $\beta_0 + \beta_1\beta_0$, then to
  - $\beta_0 + \beta_1\beta_0 + \beta_1^2\beta_0$;
- the expression involving past presidential approval, PresApp_{t-1} changed from
  - $\beta_1 \text{PresApp}_{t-1}$, to
  - $\beta_1^2 \text{PresApp}_{t-2}$, to
  - $\beta_1^3 \text{PresApp}_{t-3}$;

- the expression involving the change in disposable income, CDI_t changed from
  - $\beta_2 CDI_t$, to
  - $\beta_2 CDI_t + \beta_1\beta_2 CDI_{t-1}$, to
  - $\beta_2 CDI_t + \beta_1\beta_2 CDI_{t-1} + \beta_1^2\beta_2 CDI_{t-1}$; and
- the expression involving ϵ_t changed from simply
  - ϵ_t, to
  - $\epsilon_t + \beta_1\epsilon_{t-1}$, to
  - $\epsilon_t + \beta_1\epsilon_{t-1} + \beta_1^2\epsilon_{t-2}$.

A after an infinite number of substitutions, we obtain

$$\begin{aligned}\text{PresApp}_t &= \frac{\beta_0}{1-\beta_1} + \beta_2 CDI_t + \beta_1\beta_2 CDI_{t-1} + \beta_1^2\beta_2 CDI_{t-2} + \cdots \\ &+ \epsilon_t + \beta_1\epsilon_{t-1} + \beta_1^2\epsilon_{t-2} + \cdots\end{aligned}$$

$$\text{PresApp}_t = \frac{\theta_0}{1-\beta_1} + \sum_{k=0}^{\infty}(\beta_1^k\beta_2 CDI_{t-k} + \beta_1^k\epsilon_{t-k}). \tag{2.7}$$

This equation, though, contains no mention of any past year's presidential approval. Read as a literal causal statement, it suggests that current approval is determined by income in the current year and all previous years and by past exogenous shocks.

2.2 The Lag Operator, the Difference Operator, and Lag Operator Algebra

While the method of successive substitutions utilizes only simple algebra, it can be quite laborious. Lag operator algebra achieves the same purpose with a great deal less tedium, and so is as useful to time series analysts as matrix algebra is to those studying simple linear models.

2.2.1 The Lag Operator and Some Properties

Let X_t and Z_t be sequences of numbers associated with time periods indexed by t. The lag operator, L is defined as follows.[1]

$$LX_t = X_{t-1}.$$

---

1 Some texts refer to this as the "backshift operator," B.

The convenience of the lag operator derives partly from the fact that treating it in much the same way that multiplicative constants are treated results in mathematically correct statements.[2] For example,

$$\begin{aligned} L(LX_t) &= L^2X_t \\ &= X_{t-2}. \end{aligned}$$

In general,

$$L^kX_t = X_{t-k},$$

where k is any integer.

The lag operator is commutative with the multiplication operator: if β is a multiplicative constant,

$$\begin{aligned} L(\beta X_t) &= \beta(LX_t) \\ &= \beta X_{t-1}. \end{aligned}$$

It is also distributive over the addition operator:

$$\begin{aligned} L(Z_t + X_t) &= LZ_t + LX_t \\ &= Z_{t-1} + X_{t-1}. \end{aligned}$$

2.2.2 The Difference Operator

The **difference operator**, Δ, is also used quite heavily in time series work. It is defined at each timepoint t as the difference between the value of the series at time t and the value of the series one period earlier. That is, the first difference

$$\Delta X_t \equiv X_t - X_{t-1}.$$

Using lag operator notation,

$$\begin{aligned} \Delta X_t &\equiv X_t - X_{t-1} \\ &= X_t - LX_t \\ &= (1-L)X_t. \end{aligned}$$

It is important to note that the second difference is defined not as $X_t - X_{t-2}$. Rather,

$$\begin{aligned} \Delta^2 X_t &= \Delta(\Delta X_t) \\ &= \Delta(X_t - X_{t-1}) \\ &= \Delta X_t - \Delta X_{t-1} \\ &= (X_t - X_{t-1}) - (X_{t-1} - X_{t-2}) \\ &= X_t - 2X_{t-1} + X_{t-2}. \end{aligned}$$

2 Hamilton (1994:25–29) provides particularly good coverage of the lag operator algebra used here.

Alternatively, using the lag operator,

$$\begin{aligned}\Delta^2 X_t &= (1-L)^2 X_t \\ &= (1-2L+L^2)X_t \\ &= X_t - 2X_{t-1} + X_{t-2}.\end{aligned}$$

2.3 Lag Operator Division and Infinite Series

Sometimes we will begin with an equation in Y_{t-1} or ΔY_t and need to solve for the raw series, Y_t. Toward that end, this section introduces lag operator division.

2.3.1 The Inverse of "L"

Suppose we are given this equation for Y_{t-1}

$$Y_{t-1} = \theta + \beta X_{t-2} + a_{t-1}$$

where Y_{t-1} and X_{t-2} are random variables, θ and β are constants, and a_{t-1} is a random error term, and we want to rewrite it as an expression in Y_t. Since $Y_{t-1} \equiv LY_t$, we can think of this as "undoing" the lag operator, or equivalently, dividing by the lag operator. You may remember from ordinary algebra that a function that "undoes" its counterpart is called its inverse, and multiplying a function by its inverse is equivalent to dividing by the original function. The inverse of the lag operator is the **forward operator**, and is written L^{-1}. By definition,

$$L^{-1}L \equiv 1$$

so

$$L^{-1}LY_t = Y_t.$$

Then we can solve the previous equation for Y_t by multiplying both sides by the forward operator L^{-1} as follows.

$$\begin{aligned}Y_{t-1} &= \theta + \beta X_{t-2} + a_{t-1} \\ L^{-1}Y_{t-1} &= L^{-1}(\theta + \beta X_{t-2} + a_{t-1}) \\ Y_t &= L^{-1}\theta + L^{-1}\beta X_{t-2} + L^{-1}a_{t-1} \\ &= \theta + \beta X_{t-1} + a_t.\end{aligned}$$

2.3.2 The Inverse of "$1-\beta L$"

Division to solve for Y_t when both a Y_t and Y_{t-1} are present in the equation is more complicated, but may become more intuitive by considering geometric series from introductory calculus courses. Perhaps the most famous is the geometric series

$$1+\frac{1}{2}+\frac{1}{4}+\frac{1}{8}+\cdots+\left(\frac{1}{2}\right)^k+\cdots,$$

which has a finite sum equal to 2 even though it has an infinite number of terms. More generally,

$$\lim_{k\to\infty}\left(1+\beta+\beta^2+\beta^3+\cdots+\beta^k\right)=\frac{1}{1-\beta}$$

as long as $|\beta|<1$.

The analogue in lag operator algebra is

$$\lim_{k\to\infty}(1+\beta L+\beta^2L^2+\beta^3L^3+\cdots+\beta^kL^k)=\frac{1}{1-\beta L} \tag{2.8}$$

$$\frac{1}{1-\beta L}=\sum_{k=0}^{\infty}\beta^kL^k. \tag{2.9}$$

That is, the inverse of $1-\beta L$ can be written as the infinite sum $1+\beta L+\beta^2L^2 +\beta^3L^3+\ldots$

Suppose, for example,

$$Y_t=\beta Y_{t-1}+a_t$$

where Y_t is a random variable, β is a constant and a_t is a random error term.

We can use lag operator algebra, including Equation (2.9), to solve for Y_t as follows.

$$\begin{aligned}
Y_t &= \beta Y_{t-1}+a_t\\
Y_t-\beta Y_{t-1} &= a_t\\
(1-\beta L)Y &= a_t\\
Y_t &= \frac{1}{1-\beta L}a_t\\
&=(1+\beta L+\beta^2L^2+\beta^3L^3+\ldots)a_t\\
&=a_t+\beta a_{t-1}+\beta^2a_{t-2}+\beta^3a_{t-3}+\ldots
\end{aligned}$$

Notice that a special case of Equation (2.9) is equivalent to dividing by the difference operator. If $\beta=1$, the identity becomes

$$\frac{1}{1-L}\equiv 1+L+L^2+L^3+\ldots \tag{2.10}$$

$$=\sum_{k=0}^{\infty}L^k \tag{2.11}$$

$1-L$, though, is just lag operator notation for the difference operator; that is, $\Delta Y_t \equiv (1-L)Y_t$.

To solve an equation such as

$$\Delta Y_t = a_t$$

for Y_t, then, entails dividing by $1-L$:

$$\begin{aligned}(1-L)Y_t &= a_t\\ Y_t &= \frac{a_t}{1-L}\\ &= a_t\sum_{k=0}^{\infty}L^k\\ &= (1+L+L^2+L^3+\ldots)a_t\\ &= a_t+a_{t-1}+a_{t-2}+a_{t-3}+\ldots\end{aligned}$$

Our solution for Y_t is an infinite series involving lags of a_t.

2.4 Lag Operator Algebra: An Example

Now consider the example introduced in the beginning of this chapter. The mathematical equivalence of the two presidential approval models is easily shown by means of a derivation using lag operators.

Beginning with

$$\begin{aligned}\text{PresApp}_t &= \beta_0+\beta_1\text{PresApp}_{t-1}+\beta_2\text{CDI}_t+\epsilon_t\\ \text{PresApp}_t-\beta_1\text{PresApp}_{t-1} &= \beta_0+\beta_2\text{CDI}_t+\epsilon_t\\ (1-\beta_1L)\text{PresApp}_t &= \beta_0+\beta_2\text{CDI}_t+\epsilon_t.\end{aligned}$$

Dividing both sides by $(1-\beta_1L)$, and using what we learned about lag operators and geometric series in Equation (2.9):

$$\begin{aligned}\text{PresApp}_t &= \frac{\beta_0}{1-\beta_1L}+\frac{\beta_2\text{CDI}_t}{1-\beta_1L}+\frac{\epsilon_t}{1-\beta_1L}\\ &= \frac{\beta_0}{1-\beta_1}+(1+\beta_1L+\beta_1^2L^2+\beta_1^3L^3+\cdots)\beta_2\text{CDI}_t\\ &\quad+(1+\beta_1L+\beta_1^2L^2+\beta_1^3L^3+\cdots)\epsilon_t\\ &= \frac{\beta_0}{1-\beta_1}+\sum_{k=0}^{\infty}\beta_1^k\beta_2\text{CDI}_{t-k}+\sum_{k=0}^{\infty}\beta_1^k\epsilon_{t-k}.\end{aligned}$$

2.5 An Aside on Linear Difference Equations

Equation (2.1) is an example of a "linear difference equation," and the derivation in Section 2.1 illustrates one standard method of "solving" linear difference equations. In many time series texts, this topic receives more extensive coverage because solutions to linear difference equations provide information about the behavior of the series. In this text, lag operator algebra is typically used as an alternative, but this section introduces some basic difference equation terminology.

First, equations of the form

$$Y_t = a_0 + \sum_{i=1}^{k} a_k Y_{t-k} + X_t,$$

where Y_t and X_t are variables observed at time periods indexed by t and the a_i are constant terms are difference equations – **linear difference equations** in this case because no Y_{t-i} is squared, cubed, or raised to any power other than one. X_t is sometimes referred to as the **forcing function**. The **order** of the equation is given by the highest value of k for which a_k is non-zero.

- $Y_t = 1 + 4Y_{t-1} + 2X_t$ is a first-order difference equation.
- $Y_t = 3 + Y_{t-1} - Y_{t-2} + X_t$ is a second-order difference equation.
- $Y_t = 3 - Y_{t-2} + X_t$ is also a second-order difference equation.

A **solution** to a linear difference equation expresses Y_t as a function only of X_t and possibly a constant term representing an initial condition. Equation (2.7) is a solution to the linear difference equation (2.1) because it expresses C_t only as a function of Y_t and ϵ_t; no lags of C_t appear in the solution.

In social science applications, we rarely (if ever) work with difference equations of order higher than two, and it is simple enough to assess the properties of the series without solving difference equations.

Reference

Hamilton J.D. (1954). *Time Series Analysis*. Princeton: Princeton University Press.

3

Properties of Time Series: Mean and Variance Stationarity

Consider a regression of military expenditures in Jordan, in millions of current US dollars, on the urban population in Fiji, in thousands, during the years 1960–2019.[1] The OLS estimation results are presented in Equation (3.1), with standard errors of the estimates in parentheses.

$$\text{JorMilEx} = \underset{(107)}{-688} + \underset{(0.324)}{4.22}\ \text{FijiUrbPop} + e_t \tag{3.1}$$
$$n = 60$$
$$R^2 = 0.74.$$

The estimated coefficient of 4.22 indicates that for each thousand-person increase in the urban population of Fiji, military expenditure in Jordan increases by \$4.22 million. The estimated coefficient is statistically significant at the 0.001 level, and a conventional interpretation of the R^2 indicates that 74% of the variance in Jordan's military expenditures is explained by variation in the size of Fiji's urban population.

The suggestion of a causal relationship, though, defies credulity. As G. Udny Yule warned in his 1925 Presidential Address to the Royal Statistical Society,

> It is fairly familiar knowledge that we sometimes obtain between quantities varying with the time (time-variables) quite high correlations to which we cannot attach any physical significance whatever, although

1 Both series are part of the *World Development Indicators*, from https://datatopics.worldbank.org/world-development-indicators/ downloaded December 12, 2020. The data fields used were MS.MIL.XPND.CD and SP.URB.TOTL.

Applied Time Series Analysis for the Social Sciences: Specification, Estimation, and Inference, First Edition. Regina M. Baker.

Companion website: www.wiley.com/go/baker

> under the ordinary test the correlation would be held to be certainly "significant." As the occurrence of such "nonsense-correlations" makes one mistrust the serious arguments that are sometimes put forward on the basis of correlations between time-series ... it is important to clear up the problem how they arise ... (Yule, 1926:2)

In the case at hand, the high correlation between Jordanian military expenditures and the size of the urban population in Fiji is explained by the fact that both series are increasing over time, as shown in Figure 3.1.

In fact, any two series that trend over time will be correlated – positively, if both trend in the same direction, and negatively if one trends upward and the other trends downward. Those who have calculated correlation coefficients by hand will see this quickly by examining a partial listing of the data in Table 3.1.[2]

These are special cases of the general circumstance where a spurious relationship between two time series variables is a consequence of their sharing at least one time series property. While different methods address non-stationarity differently, taking it into account is critical to good time series modeling. Just as regression analysts are encouraged to examine scatterplots to make

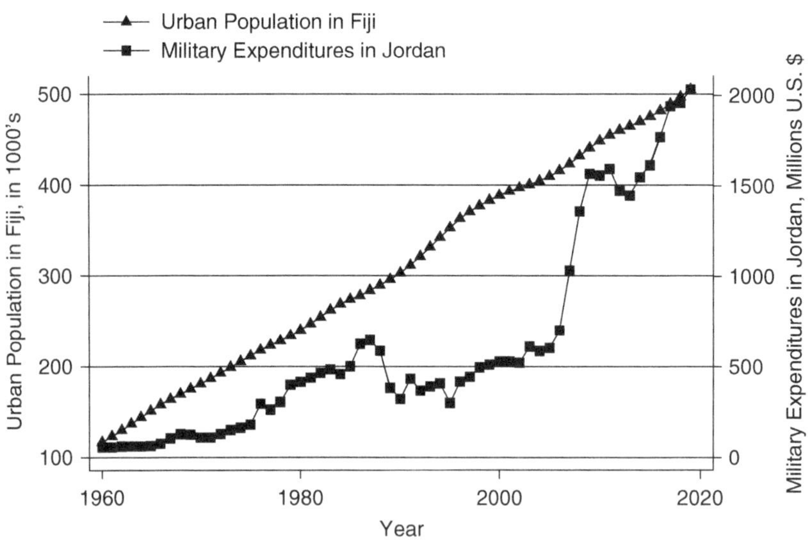

Figure 3.1 Jordanian military expenditures and Fiji's urban population, 1960–2019.

2 The high statistical significance of the relationship is magnified by properties of the OLS estimator. This is discussed further in Chapters 8 and 10.

Table 3.1 Data from a spurious regression example.

Year	Jordan's Military Expenditures (in millions)	Fiji's Urban Population (in thousands)
1960	56.1	117
1961	55.5	123
1962	60.4	130
1963	61.9	137
1964	61.9	144
⋮	⋮	⋮
2015	1615	475
2016	1769	481
2017	1940	489
2018	1958	497
2019	2032	505

preliminary decisions about model specification, time series analysts must explore the dynamic properties of their series to make preliminary decisions about model specifications. In this chapter, we will examine the properties associated with **stationarity**, tests for distinguishing between types of stationarity, and the transformations required to achieve stationarity. Apart from the intrinsic importance of choosing the correct method to transform the series, the process is a good illustration of why it is important to make choices about the model specification based on observed time series properties in the data, instead of mechanically writing a theory-based equation and estimating its parameters.

The next section of this chapter will introduce the concept of stationarity more formally, and the section following that will address two different sources of mean non-stationarity, stochastic trends and deterministic trends. Dickey–Fuller and Phillips–Perron tests to distinguish between these are presented in later sections of this chapter.

Finally, a second component of stationarity concerns the variance of time series. In the case of stochastic trends, the source of mean non-stationarity will also cause variance non-stationarity, and so transformations that address the mean non-stationarity will address both problems. Variance non-stationarity is also typical of count data, however. This topic is discussed toward the end of the chapter.

3.1 Stationarity: Formal Definitions

Thus far, we have been using the term "time series" to refer to a set of data points that can be ordered by time. In Chapter 1, the value of US Exports was one such series. In technical terms, these data represent one particular **realization** of a **data-generating process (DGP),** where the terms "realization" and "data-generating process" (DGP) are analogous to the terms "sample" and "population" when working with cross-sectional data. In the case of US Exports, the sequence (55 391, 67 491, 63 319, ... 153 409) was the particular realization of the set of random variables (Y_{1950}, Y_{1951}, Y_{1952}, ... Y_{2007}). History might have unfolded differently, but the true (unknown) DGP and the particular random events that occurred in each of the years 1950, 1951, 1952, ... 2007 resulted in the series we observe. In the same way that we use information from samples to make inferences about populations, we will use information from particular realizations to make inferences about underlying DGPs.

For the sake of convenience, we will use the shorthand Y_t to denote the ordered set of random variables $Y_1, Y_2, Y_3 \ldots Y_T$. The series Y_t is said to be **mean-stationary** if its expected value is constant over time:

$$E(Y_1) = E(Y_2) = E(Y_3) = \cdots = E(Y_T) = \mu. \tag{3.2}$$

Mean non-stationarity is often visible to the eye; contrast the time series plots in the left and right panels of Figure 3.2. While the series on the left appears to fluctuate around a constant level, the series on the right fluctuates around a level that is clearly increasing over time.

To be **weakly stationary**, a series must meet two additional criteria: its expected variance must be constant over time, and the pattern of autocorrelation in the data must be constant over time. Formally,

$$E(Y_1 - \mu)^2 = E(Y_2 - \mu)^2 = E(Y_3 - \mu)^2 = \cdots = E(Y_T - \mu)^2 = \sigma^2 \tag{3.3}$$

and

$$\text{Cov}(Y_1, Y_{1+k}) = \text{Cov}(Y_2, Y_{2+k}) = \ldots = \text{Cov}(Y_{n-k}, Y_n), \tag{3.4}$$

where k is an integer.

If in addition to these properties, the series follows a joint-normal distribution, it is *strictly stationary*. Because the assumption of weak non-stationarity is important to many widely-used models, we will examine criteria and tests associated with weak stationarity in this and the next chapter.

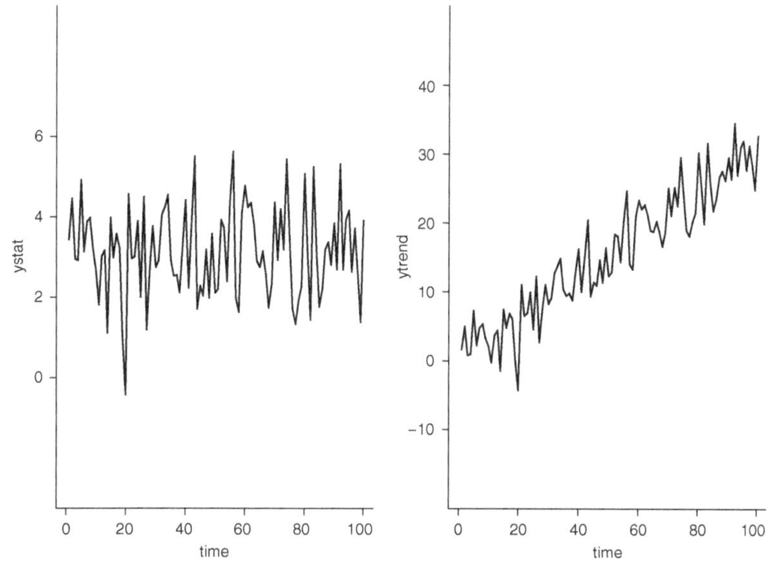

Figure 3.2 Mean stationarity versus mean nonstationarity.

3.2 Mean Non-stationarity: Stochastic Trend Versus Deterministic Trend

An intuitive strategy for modeling a series which appears to be mean-non-stationary is to include a "time" marker as an explanatory variable – for example, to estimate the regression

$$Y_t = \alpha + \beta_1 X_t + \beta_2 t + \epsilon_t \tag{3.5}$$

where t represents time periods, instead of simply estimating a bivariate regression of Y_t on X_t. This may be an appropriate method for modeling a **deterministic trend** when the source of the trend cannot be modeled directly.[3]

While this was once common practice, it is often a serious error. It is much more common to find that the trend is driven by a process like the following:

3 In economic data, the source of deterministic trends is often changing population size. If this is the case, including population as an independent variable is a better modeling strategy than adding a time variable, t: remember that t is simply a proxy, and it is preferable to use the correct variable when possible.

$$Y_t = Y_{t-1} + \theta + a_t, \tag{3.6}$$

where θ is a constant term and a_t is random error. Because there is only a one-period lag between the dependent variable and its lagged regressor, the implied coefficient of "1" on the lagged dependent variable makes this a **unit root process**. While particular realizations from the series may look the same, the two DGP's have distinct properties and should be modeled differently. Mistakenly estimating Equation (3.5) when the DGP is a unit root process is problematic for three reasons.

1) Because these trend lines are estimated by minimizing the sum of squared errors, they are highly sensitive to outliers. They are also heavily influenced by the first and last points in the series. (Remember what you learned about leverage in your introductory regression course.)
2) If θ in the stochastic trend (unit root) process is non-zero, the underlying trends of the two series are the same. The variances of the two processes are very different, however, so standard errors will be estimated incorrectly. Since the variance of a unit root process increases over time, variances of forecasts from a unit root process that is modeled as a deterministic trend instead will be understated, as we will see in later sections.
3) If $\theta = 0$ in the unit root process, many realizations of the series will appear to trend even though the underlying DGP is mean stationary.

The importance of this last point cannot be overstated. Rephrasing a bit, many series that appear to trend over time are driven by underlying processes whose expected value is constant over time. This distinction between the underlying DGP and a given realization is critical to the proper analysis of time series. Just as we recognize that it is possible to draw an atypical sample from a population, we must consider the possibility that the underlying DGP of a series is mean-stationary even if the particular realization we observe is not. To gain some intuition into why this happens, we will examine a sum generated through a series of coin tosses.

3.2.1 A Simple Illustration of Stochastic Trend: A Coin Flip

Suppose for a series of coin tosses, we create a sum by adding 1 each time heads is up and subtracting 1 for each tails. We could represent the sum this way:

$$Y_t = Y_{t-1} + h_t, \tag{3.7}$$

where

$$h_t = \begin{cases} 1 & \text{if heads} \\ -1 & \text{if tails} \end{cases}$$

and Y_t is the sum of the +1's and −1's through toss number t.

Assuming it is a fair coin, our best guess is that the number of heads we observe will be equal to the number of tails, and so the +1's and −1's will, on average, cancel each other out. Then the expected value of the sum will be 0:

$$E(Y_t) = 0.$$

Rephrased in terms of time series properties, if we plotted a series representing the sum of 1's and −1's as we continued to flip the coin, we would expect the series to be mean-stationary and fluctuate around 0.

This need not be true of particular realizations of the process, though, and in fact, a substantial proportion will appear to be mean non-stationary. Figure 3.3 shows time series plots for ten different simulations of the coin flip.

In each case, the realized Y_t is plotted on the vertical axis, with t shown on the horizontal axis.

Notice that more than half of the realizations appear to be mean-non-stationary, despite the fact that the underlying process fluctuates around a constant mean of 0. This happens because while we cannot predict a disproportionate number of heads or a disproportionate number of tails – i.e., they do not occur in expectation – they do occur in practice. Once a disproportionate number of heads has been tossed, though, there is no reason to expect a disproportionate number of tails to follow.[4] The cumulative sum is expected to remain above 0 if a disproportionate number of heads was tossed, or below 0 if a disproportionate number of tails was tossed.

Mathematically, the mean of the series Y_t is 0 in expectation. Consider the expected values of the first few realizations, though.

- At $t = 0$, there is no realization. By construction, we will start at 0.
- At $t = 1$, h_1 will be 1 if heads is tossed, and −1 otherwise. The $E(h_1) = 0$.
- At $t = 2$, h_2 will be 1 if heads is tossed, and −1 otherwise. Then $E(h_2) = 0$. Now our sum is $h_1 + h_2$. Since the expected value of both h_1 and h_2 is 0, the $E(Y_1) = 0$.

While the expected value of each h_t is 0, since there is equal probability of flipping heads or tails, each h_t is either a 1 or a −1, depending on whether heads or tails is flipped. If a disproportionate number of heads is tossed, more 1 than −1 will be added, and the $\sum_{t=1}^{t} h_t$ will exceed 0. If, on the other hand, a disproportionate number of tails is tossed, more −1 than 1 will be added, and the $\sum_{t=1}^{t} h_t$ will be less than 0.

4 Believing otherwise is known as the "Gambler's Fallacy."

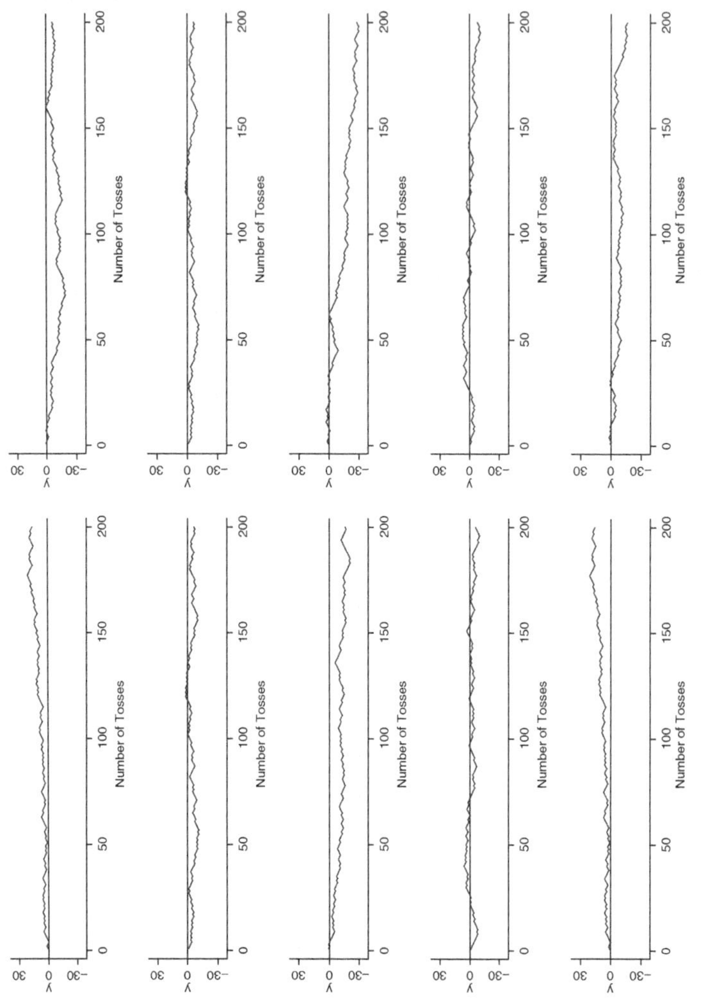

Figure 3.3 Results of ten coin toss simulations: *Yt* = heads – tails..

It should be clear that representing the observed upward or downward movement in these simulations of coin tosses by using a time trend – i.e., modeling the series as in Equation (3.8) – would be a mistake.

$$Y_t = \alpha + \beta t + e_t. \tag{3.8}$$

What appear to be trends in the coin toss experiments are due to random events, randomly occurring stretches where a disproportionate number of heads (or tails) is flipped. They are stochastic trends as opposed to deterministic trends. In the case of the coin toss, the underlying DGP is mean-stationary, but we will see in a later section that the variance increases over time.

3.2.2 Modeling Mean Non-stationarity: Deterministic Trends, Stochastic Trends, and Unit Root Processes

Discerning between deterministic trends and stochastic trends is critical to modeling them correctly. While it is always best to model the specific process driving a trend – for example, including a variable representing population size if that is what is driving the trend, deterministic linear trends can usually be modeled by including a time marker in the model, as in Equations (3.5) and (3.8).[5] Stochastic trends, on the other hand, are appropriately modeled by working with the differenced series.

The coin toss experiment is an example of a stochastic trend commonly referred to as a **random walk**, and is one of three types of **unit root** processes. In this case, the DGP was mean-stationary even though many realizations will appear to trend and the variance is non-stationary – it will increase over time. The other two types of unit root processes are non-stationary in both mean and variance. These forms, and the appropriate transformation(s) for achieving stationarity, are discussed next. Later in the chapter, we will use Dickey–Fuller and Phillips–Perron tests to evaluate whether non-stationarity in particular realizations reflects stochastic trends or deterministic trends in the DGPs.

We used Equation (3.7) to represent the sum generated by our coin flip example:

$$Y_t = Y_{t-1} + h_t,$$

where h_t is a binary random component taking the value "+1" if heads is flipped and "−1" if tails is tossed.

Using the lexicon introduced in Chapter 2, it is a first-order difference equation because it includes a one-period lag of the dependent variable, and no greater lags. This, and the implied coefficient of "1" on Y_{t-1} renders it a unit root process. Certain time series properties of unit root processes – e.g., whether they are mean-stationary or have a stable variance over time – depend on whether or not there is a constant term and/or a time trend in the DGP. In the coin flip illustration, the random component h_t is binary: for a particular

5 These were both examples of linear deterministic trends. Polynomials in the time marker can be used to model nonlinear deterministic trends.

realization, it will be either 1 or –1. When modeling time series more generally, we allow the random component to assume an infinite range of values and specify it as $a_t \sim N(0, \sigma^2)$, where σ^2 represents a constant variance. We explore three possibilities in the sections that follow.

3.2.2.1 A Random Walk Without Drift

The simplest form of a random walk or unit root process is analogous to our coin flip example and presented in Equation (3.9).

$$Y_t = Y_{t-1} + a_t, \tag{3.9}$$

where

- a_t is a random ("white noise") error term,
- $a_t \sim N(0, \sigma^2)$, and
- σ^2 is a constant.

This unit root process has neither a constant term nor a deterministic time trend, in contrast to the other possibilities. To explore how the mean and variance of the process unfold over time, we will use the method of successive substitutions to solve the linear difference equation. Suppose we choose a starting point, Y_0. Then

$$Y_1 = Y_0 + a_1,$$

and we can derive equations for successive values of Y_t by substitution:

$$\begin{aligned} Y_2 &= Y_1 + a_2 \\ &= Y_0 + a_1 + a_2 \end{aligned}$$

$$\begin{aligned} Y_3 &= Y_2 + a_3 \\ &= Y_0 + a_1 + a_2 + a_3. \end{aligned}$$

A pattern quickly emerges:

$$Y_t = Y_0 + \sum_{k=1}^{t} a_k. \tag{3.10}$$

We can use Equation (3.10) to examine how the mean and variance of the series change over time.[6] The mean of the series is given by its expected value,

$$\begin{aligned} E(Y_t) &= E(Y_0 + \sum_{k=1}^{t} a_t) \\ &= E(Y_0 + a_1 + a_2 + \cdots + a_t) \\ &= E(Y_0) + E(a_1) + E(a_2) + \cdots + E(a_t). \end{aligned}$$

6 Appendix 1 includes a review of expected value calculations that may be helpful to readers unfamiliar with them.

Remember that a_t is a random white noise process with mean 0 by construction, so the $E(a_t)$ terms can be dropped. Then for our simplest unit root process, represented by Equation (3.9),

$$E(Y_t) = Y_0.$$

The expected value of the series is constant over time, so the series is mean-stationary.

The expected variance of the series, however, is not constant over time. Consider the variance of Equation (3.9) for successive time points.

$$\begin{aligned}\mathrm{Var}(Y_1) &= E[Y_1 - E(Y_1)]^2 \\ &= E[(Y_0 + a_1 - Y_0)^2] \\ &= E(a_1^2).\end{aligned}$$

Again, though, recall that $E(a_t) = 0$ by construction. Then we can further simplify:

$$\begin{aligned}\mathrm{Var}(Y_1) &= E(a_1 - 0)^2 \\ &= \mathrm{Var}(a_1) \\ &= \sigma^2\end{aligned}$$

At $t = 2$,

$$\begin{aligned}\mathrm{Var}(Y_2) &= E[Y_2 - E(Y_2)]^2 \\ &= E[(Y_0 + a_1 + a_2 - Y_0)^2] \\ &= E[(a_1 + a_2)^2] \\ &= E[a_1^2 + 2a_1a_2 + a_2^2] \\ &= E(a_1^2) + E(a_1a_2) + E(a_2^2).\end{aligned}$$

Now recall the assumptions that the random error terms are independent of each other: $E(a_1a_2) = 0$.

Also by assumption, $E(a_1^2) = E(a_2^2) = \sigma^2$. Then

$$\mathrm{Var}(Y_2) = 2\sigma^2.$$

Using the same method to derive the variance of Y_3, you will find that

$$\mathrm{Var}(Y_3) = 3\sigma^2.$$

In general terms,

$$\mathrm{Var}(Y_t) = t\sigma^2.$$

The variance of this series is increasing over time, so it is not variance-stationary, and may need to be transformed. In the case of unit root processes, differencing the series will address the variance non-stationarity.

3.2.2.2 A Random Walk with Drift

Unlike the simplest form of the unit root process, the other two forms are both mean and variance non-stationary. The second type takes the form represented by Equation (3.11).

$$Y_t = Y_{t-1} + \theta_0 + a_t, \tag{3.11}$$

where θ_0 is a constant term, often referred to as a "drift term."[7]

To see that the mean and variance of this series are both non-stationary, we once more define an arbitrary starting point Y_0 and will use lag operator algebra to solve the difference equation. To be consistent with the general equation, we will stipulate that $Y_0 = \theta_0$.

$$Y_t = Y_{t-1} + \theta_0 + a_t$$

$$\begin{aligned} Y_t - Y_{t-1} &= \theta_0 + a_t \\ (1-L)Y_t &= \theta_0 + a_t \\ Y_t &= \frac{\theta_0}{1-L} + \frac{a_t}{1-L}. \end{aligned}$$

Recalling the properties of lag operators and geometric series from Section (2.3),

$$\begin{aligned} Y_t &= \frac{\theta_0}{1-L} + \frac{a_t}{1-L} \\ &= \sum_{k=0}^{t} L^k \theta_0 + \sum_{k=0}^{t} L^k a_t \end{aligned}$$

Since θ_0 is a constant,

$$L\theta_0 = L^2\theta_0 = \cdots = L^t\theta_0 = \theta_0.$$

Then

$$\begin{aligned} Y_t &= (t+1)\theta_0 + (1 + L + L^2 + L^3 + \cdots + L^t)a_t \\ &= \theta_0 + \theta_0 t + (a_t + a_{t-1} + a_{t-2} + a_{t-3} + \cdots + a_1) \end{aligned}$$

7 The term "drift" is used in two distinct ways in the literature. In older literature, "drift" often refers to the meandering that unit root processes exhibit whether or not the DGP includes a constant ("drift") term. More recent literature is more likely to include references to the "drift term," meaning a constant term in a unit root process.

Recalling our initial condition, $Y_0 = \theta_0$, we can write the general equation

$$Y_t = Y_0 + \theta_0 t + \sum_{k=1}^{t} a_k. \tag{3.12}$$

Notice that the presence of the constant term θ_0 in this second type of unit root process has resulted in an equation that includes a time trend; the series will increase over time if $\theta_0 > 0$ and decrease over time if $\theta_0 < 0$. We can derive this result more formally by examining the expected value of the series:

$$E(Y_t) = E(Y_0 + \theta_0 t + \sum_{k=1}^{t} a_t).$$

As in the previous case, the $E(a_t)$ terms will drop out since $a_t = 0$ in expectation. We are left with

$$E(Y_t) = Y_0 + \theta_0 t,$$

a simple equation describing a series whose expected value is a linear function of time.

This is what is meant by **stochastic trend**[8] rather than deterministic trend. In these cases, Equation (3.8) does not accurately reflect the DGP. Furthermore, attempting to approximate the stochastic trend with a deterministic trend is unsatisfactory for two additional reasons. First, it does not address the variance non-stationarity associated with the stochastic trend. Unit root processes always have variances that increase over time. Second, as we saw earlier in the chapter, deterministic trend estimations are heavily influenced by the endpoints of the series; they are also heavily influenced by outliers. Differencing the series yields both a better representation of the DGP and a transformation that weights each data point evenly.

3.2.2.3 A Random Walk with Drift and Trend

The most complicated of the three types of random walk processes is represented by Equation (3.13). In this case, the series includes a random walk component, a drift (or constant) term, and a deterministic time trend.

$$Y_t = Y_{t-1} + \theta_0 + \beta t + a_t \tag{3.13}$$

$$Y_1 = Y_0 + \theta_0 + \beta + a_1$$

---

8 "Stochastic trend" is, at first glance, an unfortunate oxymoron, but it is standard parlance in the time series literature.

$$\begin{aligned} Y_2 &= Y_1 + \theta_0 + 2\beta + a_2 \\ &= Y_0 + \theta_0 + \beta + a_1 + \theta_0 + 2\beta + a_2 \\ &= Y_0 + 2\theta_0 + 3\beta + a_1 + a_2 \end{aligned}$$

$$\begin{aligned} Y_3 &= Y_2 + \theta_0 + 3\beta + a_3 \\ &= Y_0 + 2\theta_0 + 3\beta + a_1 + a_2 + \theta_0 + 3\beta + a_3 \\ &= Y_0 + 3\theta_0 + 6\beta + a_1 + a_2 + a_3. \end{aligned}$$

The general form is a little more difficult to see, but it is easily verified that it can be written as

$$Y_t = Y_0 + \theta_0 t + \beta \frac{t(t+1)}{2} + \sum_{k=1}^{t} a_k. \tag{3.14}$$

In this case, there are two sources of mean non-stationarity – the $\theta_0 t$ term and the $\beta \frac{t(t+1)}{2}$ term. As with the other two cases, the variance increases with time because of the accumulation of random error terms.

All three of these cases have variance non-stationarity that is appropriately addressed by differencing the series. The latter two are also mean non-stationary. With practice, it becomes easier to make good guesses about whether mean non-stationarity is due to a stochastic trend or a deterministic trend simply by looking at a graph; the variance in series with stochastic trends results in the series drifting away from its trend line and then returning. Dickey–Fuller tests provide a formal means.

3.3 Dickey–Fuller (D–F) Tests

The chief purpose of D–F tests is to decide whether there is non-stationarity in our series consistent with either a unit root process or with the pure trend model shown in Equation (3.15) below.

$$Y_t = \theta_0 + \beta t + a_t. \tag{3.15}$$

An intuitive but incorrect approach to testing for unit roots would be to estimate a version of Equation (3.13), which encompasses the stochastic trend and deterministic trend cases, generalizing it by replacing the implied coefficient of "1" on the Y_{t-1} term with a parameter ρ, as in (3.16).

$$Y_t = \rho Y_{t-1} + \theta_0 + \beta t + a_t. \tag{3.16}$$

If even this much were appropriate, we might then choose the strongest candidate for the DGP by testing hypotheses concerning ρ, θ_0, and β.

In the case of a unit root, though, this model cannot be estimated directly. Additionally, a combination of unusual features of D–F tests must be considered.

- If $\rho = 1$ as in the unit root case, the variance of the series increases over time. Not only would the estimated variances used in significance tests be incorrect, but parameter estimates would be biased. To address this, D–F tests are based on modified versions of our unit root equations.
- The null hypothesis is that there is a unit root, and the tests have low power, leading to concerns that we might infer the presence of a unit root when there isn't one. Because the power of the unit root test increases when we drop the trend term, it's not uncommon to reject the null hypothesis based on estimates from restricted models such as Equations (3.9) and (3.11) but fail to reject based on estimates from the full model, Equation (3.13).
- Dickey and Fuller provided two sets of relevant distributions, which raises questions about the best testing strategy. While the t-statistic associated with the unit root component of the correctly-specified D–F equation is used to evaluate the null hypothesis that $\rho = 1$, it follows a special distribution. In addition to these distributions (one for each of the three unit root models), Dickey and Fuller provided special distributions for testing joint hypotheses – e.g. $\rho = 1$ and $\theta_0 \neq 0$ and $\beta = 0$ – using F-tests.

These concerns are explained in turn in the next sections of this chapter. Combined with the multiple possibilities presented by different estimating equations and the choice between t- and F-tests , they provide the basis for the Elder–Kennedy strategy for unit root testing which follows.

3.3.1 D–F Estimating Equations and the D–F Distributions

Although Equation (3.16) and the analogues associated with Equations (3.9) and (3.11) cannot be estimated directly, transformed versions can. We begin by subtracting Y_{t-1} from both sides of Equation (3.16), and then rearrange terms as follows.

$$Y_t - Y_{t-1} = \rho Y_{t-1} - Y_{t-1} + \theta_0 + \beta t + a_t$$
$$\Delta Y_t = (\rho - 1)Y_{t-1} + \theta_0 + \beta t + a_t.$$

The D–F test associated with the unrestricted equation, then, is performed by estimating Equation (3.17),

$$\Delta Y_t = \alpha Y_{t-1} + \theta_0 + \beta t + a_t \tag{3.17}$$

where $\alpha \equiv \rho - 1$ and $a_t \sim N(0, \sigma^2)$.

The analogous D–F estimating equations for the other two unit root models are presented in the second column of Table 3.2. The additional columns will be explained later in the chapter.

Assuming ΔY_t is stationary, these equations can be estimated using OLS. The test statistic is $\alpha / SE(\alpha)$, but it does not follow the t-distribution and so the t-test included with ordinary regression output should not be used for the unit root test. Dickey and Fuller derived asymptotic distributions $\tau_1 - \tau_3$, one for each of the three unit root models, and these must be used instead. More recently, MacKinnon derived approximate distributions that can be used in small samples as well.

3.3.2 Warning: the Null Hypothesis and Low Power

Tests in this family – Dickey–Fuller tests, Augmented Dickey–Fuller tests and Phillips–Perron tests – share an unusual feature: the null hypothesis is that there *is* a unit root:

$$H_o:\ \alpha = 0,$$

$$H_a:\ \alpha < 0.$$

Table 3.2 Three unit root models and associated D–F tests.

Unit Root Model	Series Properties	Associated D–F Model	Augmented D–F Model
$Y_t = Y_{t-1} + a_t$	Mean Stationary	$\Delta Y_t = \alpha Y_{t-1} + a_t$	$\Delta Y_t = \alpha Y_{t-1} + a_t + \sum_{i=1}^{p} \gamma_i \Delta Y_{t-i}$
	Variance grows over time	where $\alpha = \rho - 1$	
$Y_t = Y_{t-1} + \theta_0 + a_t$	Mean Non-Stationary	$\Delta Y_t = \alpha Y_{t-1} + \theta_0 + a_t$	$\Delta Y_t = \alpha Y_{t-1} + \theta_0 + a_t + \sum_{i=1}^{p} \gamma_i \Delta Y_{t-i}$
	Variance grows over time	where $\alpha = \rho - 1$	
$Y_t = Y_{t-1} + \beta t + \theta_0 + a_t$	Mean Explosive	$\Delta Y_t = \alpha Y_{t-1} + \beta t + \theta_0 + a_t$	$\Delta Y_t = \alpha Y_{t-1} + \beta t + \theta_0 + a_t + \sum_{i=1}^{p} \gamma_i \Delta Y_{t-i}$
	Variance grows over time	where $\alpha = \rho - 1$	

Moreover, they have low statistical power. That is, the probability of making a type II error – failing to reject the null hypothesis when the null hypothesis is false – is high relative to the power of tests associated with ordinary regression models.[9] The combination of the unusual null hypothesis and low power means that there is a substantial risk of inferring from D–F tests that there is a unit root when there isn't one. Later in the chapter we will discuss differencing as a strategy for working with unit root series, and explain how to assess whether our inference of a unit root was mistaken.

As an aside, the careful reader will notice that the test involving the t-statistic is one-sided. Although $\alpha < -2$, corresponding to $\rho < -1$ would also lead to a non-stationary series, negative autocorrelation is not possible in the continuous processes we are typically modeling.[10] This is rarely a limiting assumption, since even when working with discrete data, we're typically modeling an underlying process that is continuous, but measured only at discrete intervals.

3.3.3 Testing for Drift and Trend: Individual Coefficients (t-statistics) Versus Restricted Regression ("F-tests")

Finally, in addition to providing appropriate distributions for testing the null hypothesis that $\alpha = 0$ (equivalently, $\rho = 1$) in the three D–F estimating equations, Dickey and Fuller provided an alternative means of discerning between the three types of unit root processes. Because the method relies on the restricted regression computation that is performed with F-tests in ordinary regression, the tests are commonly referred to as "F-tests," but require special D–F distributions called $\phi 1$–$\phi 3$.

Recall that the ANOVA table provided with standard regression output includes a decomposition of the variance of a dependent variable Y_t into a piece "explained by" the independent variable(s) and a piece calculated from the residuals. We express this decomposition in terms of three sums of squares:

- the Total Sum of Squares, TSS, $\sum_{t=1}^{T}(Y_t - Y)^2$,
- the Explained Sum of Squares, ESS, $\sum_{t=1}^{T}(\hat{Y} - \bar{Y})^2$, and
- the Residual Sum of Squares, RSS, $\sum_{t=1}^{T}(Y_i - \hat{Y})^2$
- $\text{TSS} = \text{ESS} + \text{RSS}$.

9 Remember, this is distinct from the level of statistical significance, which is the probability of a type I error – rejecting the null hypothesis when the null hypothesis is true.

10 To see this, try to draw a continuous process with a negative autocorrelation.

Models with higher explanatory power have lower RSS's relative to TSS. The D–F "F-tests" are constructed as follows.

1) Estimate two models with the same dependent variable, omitting terms from one. The model with a subset of the other's explanatory variables is the "restricted" model.
2) Calculate the following test statistic from the RSS's provided in the ANOVA table.
3) Compare the test statistic to the appropriate D–F ϕ distribution.

$$\phi_i = \frac{[\text{RSS(restricted)} - \text{RSS(unrestricted)}]/r}{\text{RSS(unrestricted)}/(T-k)} \tag{3.18}$$

where

- RSS(restricted) is the RSS obtained by estimating the model that omits a term (or terms),
- RSS(unrestricted) is the RSS obtained by estimating the more general model,
- r is the number of restrictions,
- T is the number of usable observations, and
- k is the number of terms estimated in the unrestricted model.

For example, the D–F equation corresponding to the simplest unit root process

$$\Delta Y_t = \alpha Y_{t-1} + a_t \tag{3.19}$$

is a restricted version of the unit root process that includes a drift term,

$$\Delta Y_t = \alpha Y_{t-1} + \theta_0 + a_t. \tag{3.20}$$

Assuming we have ruled out the possibility of a deterministic trend, we can test the joint hypothesis that our DGP has a unit root but no drift term, $\alpha = \theta_0 = 0$, by estimating Equations (3.19) and (3.20), using the RSS's from the ANOVA table provided in the output, calculating the test statistic using Equation (3.18), and comparing the statistic to D–F's ϕ_1 distribution.

Dickey and Fuller provided two additional ϕ distributions for joint tests in involving model restrictions, both of which require estimating the most general of the three D–F estimating equations, Equation (3.17). A summary of the three possibilities is presented in Table 3.3.

While the restricted regression approach has been widely used, Dickey (1984) and Elder and Kennedy (2001a, 2001b) have shown that they are less powerful than the tests involving the t-statistics, a prominent consideration in the Elder–Kennedy testing strategy.

Table 3.3 Dickey–Fuller's restricted regression tests ("F-tests").

Equations Estimated*	Restrictions Tested	D–F Distribution
3.20: $\Delta Y_t = \alpha Y_{t-1} + \theta_0 + a_t$ 3.19: $\Delta Y_t = \alpha Y_{t-1} + a_t$	$\alpha = \theta_0 = 0$	ϕ_1
3.17: $\Delta Y_t = \alpha Y_{t-1} + \beta t + \theta_0 + a_t$ 3.19: $\Delta Y_t = \alpha Y_{t-1} + a_t$	$\alpha = \beta = \theta_0 = 0$	ϕ_2
3.17: $\Delta Y_t = \alpha Y_{t-1} + \beta t + \theta_0 + a_t$ 3.20: $\Delta Y_t = \alpha Y_{t-1} + \theta_0 + a_t$	$\alpha = \beta = 0$	ϕ_3

*The equations shown are ordinary D–F estimating equations, but the distributions can also be used with the Augmented D–F tests, explained below.

3.4 Unit Root Testing Strategies: Elder–Kennedy's Simplified Approach

In a 1981 follow-up to their original (1979) paper, Dickey and Fuller sketched a strategy that involved estimating all three D–F estimating equations, using the D–F F-tests to decide which is preferred, and then testing the null hypothesis that $\alpha = 0$ using the preferred D–F equation. Recently, however, Elder and Kennedy, 2001b have introduced a simplified strategy that addresses several concerns. Most importantly, the Elder–Kennedy approach eliminates the double– and triple-testing that raises questions about the true statistical significance of the tests. Also, it incorporates a critical revision of earlier throught. While Dickey and Fuller originally believed that their F-tests were more powerful than their t-test (1981), subsequent work indicates that this inference was driven by the particular starting values used in their simulations and infers the opposite–that the t-tests are more powerful than the F-tests. (Dickey, 1984; and Elder & Kennedy, 2001A) Their simplified strategy is based on the following insights.

- While the tests' low power sometimes leads to the uncomfortable inference that there is both a stochastic trend and a deterministic trend, this is not a realistic outcome and need not be considered.[11]
- In many cases, we need not consider all four of the remaining possibilities (the three remaining possibilities for a trending series, and the possibility of no trend).

11 See Elder & Kennedy 2001B: 145 (footnote 2), Perron, 1988:304, and Holden and Perman, 1994 (Holden and Perman, 1994:63).

- If we are sure there is a trend – i.e., the series is mean-non-stationary – then we're left with the pure trend case, Equation (3.15), or a unit root process with a drift term, as in Equation (3.11).
- On the other hand, if we are sure there is not a trend, then either the series has a unit root but no drift term, as in Equation (3.9), or it has neither a deterministic nor a stochastic trend.

- If we are not sure whether there is a trend or not, the insight that the t-tests are more powerful than the F-tests still allows us to pursue a simpler strategy.

The Elder–Kennedy strategy has two pieces. The first is appropriate for cases in which we are either sure that there is mean non-stationarity or sure that there is not mean non-stationarity. The second is appropriate if we are uncertain about whether or not there is mean non-stationarity.

3.4.1 Part I: Clear Mean Stationarity or Clear Mean Non-stationarity

In some cases we are certain a series is mean non-stationary, or we are certain that it isn't. The former is often true when we observe a clear trend in a series plot. The latter includes cases where series cannot logically increase or decrease indefinitely, such as series that are logically bounded, like percentages. The Elder–Kennedy process for unit root testing in these cases is sketched in Figure 3.4.

If we are certain there is mean non-stationarity, we have narrowed the initial choices for our DGP to Equations (3.11) and (3.15). We begin by estimating the most general of the three D–F equations, Equation (3.17), since it encompasses both possibilities, and test the null hypothesis that $\alpha = 0$ using the t-statistic and either a Dickey–Fuller distribution, if we have a long series, or MacKinnon's distribution for shorter series. If we cannot reject the null hypothesis, we model a unit root; otherwise we model a deterministic trend. In the event that the low power of the test leads us to model a unit root when a deterministic trend is appropriate, we will be able to catch the error by differencing the series (discussed later in the chapter) and checking the autocorrelation plot for a specific type of moving average component (explained in Chapter 4).

If, on the other hand, we are certain that the DGP is mean-stationary, we have narrowed the choices for our DGP to Equation (3.9) or a series without

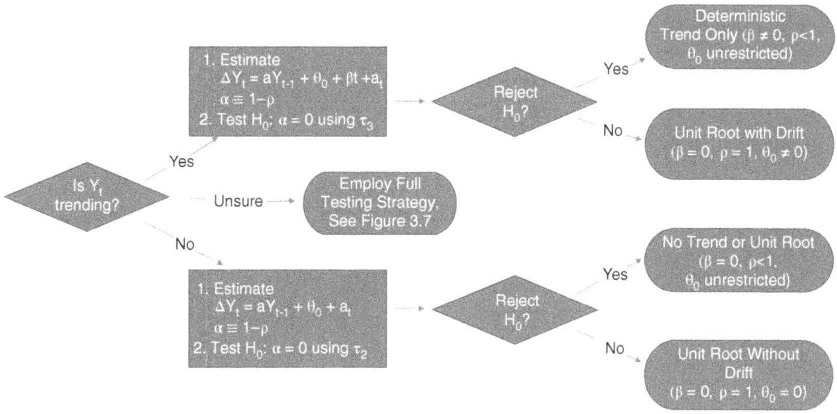

Figure 3.4 Elder–Kennedy unit root testing strategy, Part I.

either a unit root or deterministic trend. To decide between them, we estimate Equation (3.20) and test the null hypothesis that $\alpha = 0$ using the t-statistic and either Dickey–Fuller's asymptotic distribution, if we have a large number of cases, or Mackinnon's distribution for shorter series. If we cannot reject the null hypothesis, we model a unit root; otherwise we infer that there is neither a deterministic nor stochastic trend to model.

The strategy in the case of certain mean non-stationarity is illustrated using monthly data for the number of armed robberies in Boston; the strategy for series where there is no trend is illustrated using monthly data for the average monthly intravenous (IV) iron dose for US End-stage Kidney Disease (ESKD) patients.

3.4.2 Example 1: Boston Robbery Series

Dickey–Fuller tests are a relatively new development, and were not available to early time series analysts. Well-trained researchers tended to assume any series exhibiting mean non-stationarity contained a stochastic trend, and differenced the series to achieve stationarity when needed. Dickey–Fuller tests performed for the examples used in McCleary and Hay's excellent *Applied Time Series Analysis for the Social Sciences* support this decision in all cases except one, the Boston Robbery Series, and McCleary and Hay noted throughout their analysis of the differenced series that subsequent diagnostic tools did not yield clear guidance.

The series, shown in Figure 3.5, represents the number of armed robberies reported in Boston between January of 1966 and October of 1977. The

Bartley–Fox Gun Law establishing minimum sentences for illegally carrying a firearm was passed in 1974, and went into effect in April of 1975. A preliminary step toward evaluating the effect of the gun law will be diagnosing the pre-intervention properties of the series. It is clear that the series is increasing over time. The series also appears to exhibit variance non-stationarity; it is less variable at lower levels in the earlier periods than it is at higher levels during the later periods. As explained in earlier sections, variance that increases over time is a typical feature of a unit root process, so we will begin by testing for a unit root. Following the Elder–Kennedy strategy, we estimate Equation (3.17) for the pre-intervention observations. The estimated equation is

$$\Delta Y_t = \underset{(-5.51)}{-0.44} Y_{t-1} + \underset{(5.21)}{1.58} t + \underset{(-0.64)}{-4.58} + a_t, \tag{3.21}$$

where $n = 112$, $a_t \sim N(0, \sigma^2)$, and estimated t-statistics are presented in parentheses. The critical values for the Dickey–Fuller test are –3.449 and –4.037 for the 5% and 1% levels of significance respectively. Since the t-statistic associated with the unit root is –5.51, below the critical value for the 1% level of significance, we reject the null hypothesis of a unit root, $p < .01$.

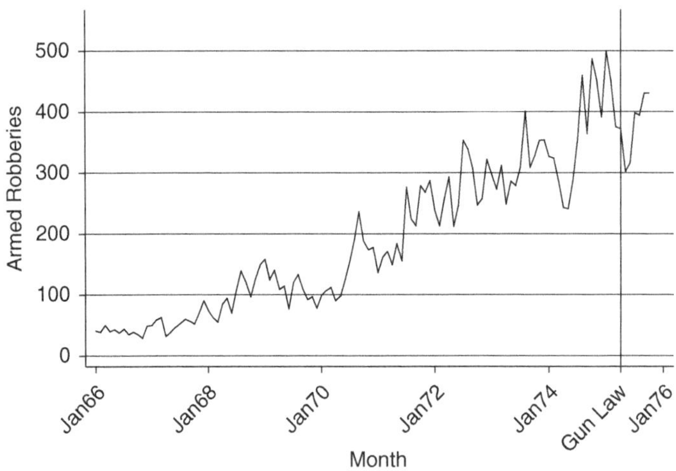

Figure 3.5 Boston Robberies, January 1966 – October 1975.

While modeling the deterministic trend addresses the mean non-stationarity, it does not address the variance non-stationarity, so we will continue with this example later in the chapter.

3.4.3 Example 2: IV Iron Among US ESKD Patients

Patients receiving regular dialysis to address ESKD are typically anemic, and many receive IV iron as a result. In the United States, treatment for ESKD is a Medicare (government-funded) entitlement regardless of patient's age. Partly in response to concerns that dialysis providers were over-prescribing and over-treating in order to boost profits, the Medicare Improvements for Patients and Provider Act of 2008 (MIPPA) mandated a payment reform beginning in January of 2011 that changed incentives to providers. By introducing a bundled payment (fixed amount per patient type), the reform eliminated the incentive to provide unneeded treatment under the old fee-for-service model.

The series shown in Figure 3.6 represents the average monthly IV iron dose among US ESKD patients receiving IV iron. In Chapter 6, we will begin discussing methods for modeling the change observed in the post-reform series. The first step, however, will be to model the pre-reform series so that it can be used as a baseline.

Although there appears to be a slight increase in the average dose between 2005 and 2011, we can rule out the possibility of a deterministic linear trend in the DGP on logical grounds: because dosages cannot fall below zero or exceed certain limits, the average monthly dose is bounded both below and above. Modeling the series as mean-stationary when we clearly see a slight trend may seem counter-intuitive, but recall our coin toss example. We know the DGP for the sum calculated by adding 1 for each "heads" toss and subtracting 1 for each "tails" toss must be mean-stationary, even if a particular realization is not. The same logic applies here.[12] Series where the variable of interest is a percentage are other common cases where linear trends in the DGP can often be eliminated from consideration, since percentages are bounded below by 0 and above by 100.

Following the Elder–Kennedy strategy, then, we estimate Equation (3.20) and obtain the following results.

$$\Delta Y_t = \underset{(-11.85)}{-1.35} Y_{t-1} + \underset{(11.85)}{498.15} + a_t, \tag{3.22}$$

12 Analysts whose primary concern is a short-term forecast may prefer to model a linear trend anyway. See Chapter 5 for further discussion.

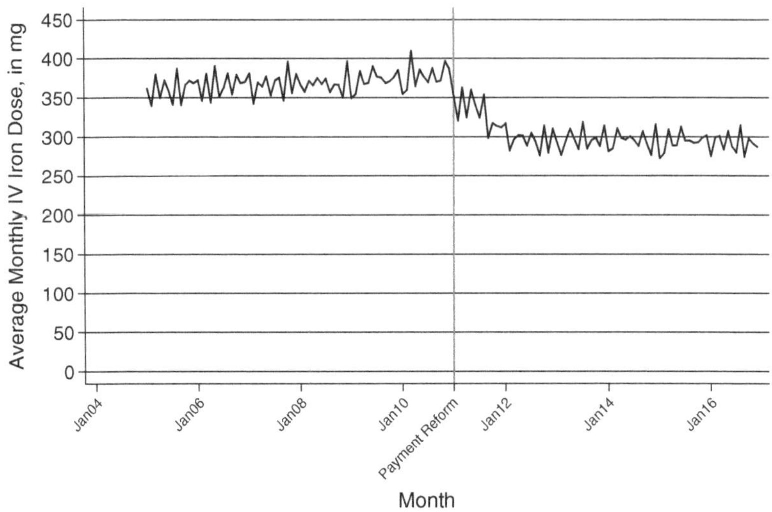

Figure 3.6 Average monthly IV iron dose (mg) among US ESKD patients receiving IV iron, 2005–2016. (Data source: United States Renal Data System, 2018 USRDS annual data report.)

where $n = 71$, $a_t \sim N(0, \sigma^2)$, and estimated t-statistics are presented in parentheses. The critical values for the Dickey–Fuller test are −2.913 and −3.551 for the 5% and 1% levels of significance respectively. Since the t-statistic associated with the unit root is −11.85, well below the critical value for the 1% level of significance, we reject the null hypothesis of a unit root, $p < .01$.[13]

3.4.4 Part II: Uncertain Mean Non-stationarity

The strategy becomes more complicated if our knowledge of the series leaves us uncertain about whether it is mean non-stationary or not. The process, sketched in Figure 3.7, entails executing the following sequence of steps.

1) Estimate the most general model, Equation (3.17), and test the null hypothesis that $\alpha = 0$ using the D–F distribution, with a sufficient number of cases, or the Mackinnon distribution for small samples.

---

13 The data reported here have been supplied by the United States Renal Data System (USRDS). The interpretation and reporting of these data are the responsibility of the author(s) and in no way should be seen as an official policy or interpretation of the US government.

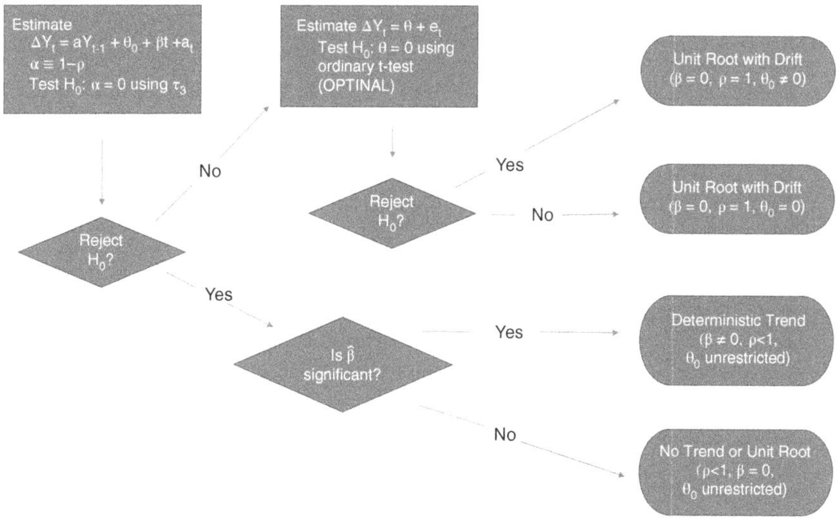

Figure 3.7 Elder–Kennedy unit root testing strategy, Part II.

2) If we reject the null hypothesis of a unit root, evaluate the significance of the trend term, β, using an ordinary t-test. As with all such evaluations, it is important to consider both the substantive significance – does $\hat{\beta}$ suggest a meaningful trend? – as well as the p-value. In shorter series, our point estimate may indicate a trend worth considering even though a test of statistical significance fails.

- If the estimated β, the coefficient on the trend term, is sizable and statistically significant according to the associated t-test, then model a deterministic trend.
- If the estimated β, the coefficient on the trend term, is sizable but not statistically significant, some judgment is required. We may want to include a deterministic trend as a control variable.
- If the estimated β is neither sizable nor statistically significant, we need not model either a unit root or a deterministic trend.

3) If we cannot reject the null hypothesis of a unit root, then we will need to difference the series to proceed.

- If it is important to decide whether or not there is a drift term – i.e., to know whether the series is both mean non-stationary and variance non-stationary or only the latter – Elder and Kennedy recommend regressing the differenced series on a null model, a model with only a constant term, and testing the null hypothesis that the constant term is 0. Rejecting this null hypothesis leads to the inference that the series has a stochastic trend; failure to reject suggests there isn't. In many applications, the presence of a unit root will be addressed by differencing the series whether or not there is a drift term, so this step involving a second regression may be unnecessary.

We will follow this process in Example 3, below.

3.4.5 Example 3: The Dow Jones Industrial Average (A U.S. Stock Index)

The Dow Jones Industrial Average (DJIA) is a stock index designed to be an indicator of the performance of the US industrial sector. It is published daily when the US stock markets are open (i.e., excluding weekends and holidays). The DJIA for January 2, 2015 through December 31, 2015 is shown in Figure 3.8.[14]

For the first several months, the series alternates between short periods of increases and short periods of decreases, with the former seeming to

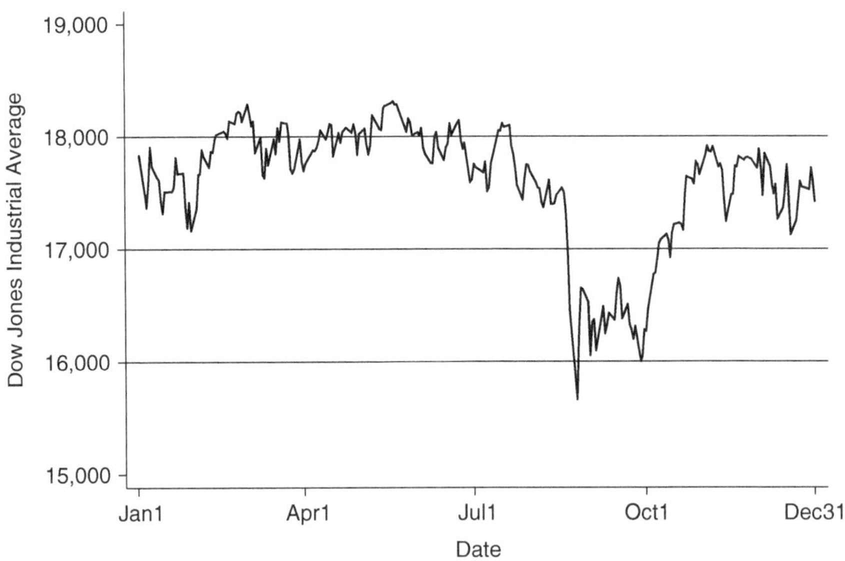

Figure 3.8 The Dow Jones Industrial Stock Index Series: January 2–December 31, 2015.

14 The Dow Jones Industrial Average© data were downloaded from http://www.djaverages.com January 24, 2016.

outweigh the latter for a possible net increase. The series includes a dramatic fall in August, though, corresponding to a stock market crash at that time, which complicates assessments of mean non-stationarity in the DGP. This pattern – a drastic shift from a relatively consistent pattern – is referred to as a structural break. In this section, we will test for a unit root ignoring the structural break. In a later section, I will introduce Phillips–Perron tests, unit root tests that take structural breaks into account, and compare the results.

The first step is to estimate the most general form, consistent with Equations (3.13) and (3.17):

$$\Delta DJIA_t = \alpha DJIA_{t-1} + \theta_0 + \beta t + a_t. \tag{3.23}$$

The results are presented in the first line of Table 3.4. Because the test statistic, –2.631, is not larger in absolute value than the critical value for the D–F distribution at the 10% level of significance, we cannot reject the null hypothesis of a unit root. Consistent with this result, the MacKinnon p-value is 0.267.

Notice the importance of using the D–F distribution instead of the ordinary t-distribution. Had we mistakenly done the latter, we would have rejected the null hypothesis of a unit root, since the estimated α is more than two standard errors away from 0; indeed, the associated p-value is 0.009.

In many cases, we will difference the series to achieve stationarity before proceeding. This is required for ARIMA modeling, but may also be preferred if we are working with ordinary regression models. The logic behind differencing unit root series is explained in the next section.

Even using methods that involve working with the raw series, though, differencing the series provides a means of doing two checks on our

Table 3.4 Dickey–Fuller tests for the DJIA series.

Series	α	θ_0	β	Test Statistic	1% Critical Value	5% Critical Value	10% Critical Value	MacKinnon P-value
Raw	−0.055	979	−0.165	−2.631	−3.990	−3.430	−3.130	0.267
	(0.009)	(0.009)	(0.300)					
Differenced	−0.999	2.667	−0.024	−15.785	−3.990	−3.430	−3.130	<0.0001
	(<0.001)	(0.606)	(0.590)					

P-values associated with ordinary t-tests are reported in parentheses.

inference of a unit root. First, if the differenced series is non-stationary, we need to apply the same strategy to the differenced series to see if we have a higher-order unit root. Second, if our inference of a unit root was mistaken and we have a deterministic trend instead – a danger given the low power of our unit root tests – we will be able to detect the problem by examining the autocorrelation function of the differenced series. The logic is explained in the next section, and the autocorrelation function is explained in Chapter 4.

3.5 Transforming Unit Root Series to Achieve Stationarity: Differencing

In methods such as ARIMA modeling, transforming a non-stationary series to stationarity is required to proceed. Regardless of our chosen method, though, differencing the suspected unit root series is important to confirm our diagnosis.

3.5.1 Differencing to Achieve Stationarity

To see why differencing a unit root series removes the non-stationarity, consider Equation (3.9), the simplest unit root process

$$Y_t = Y_{t-1} + a_t$$

and recall that it is mathematically equivalent to Equation (3.10):

$$Y_t = Y_0 + \sum_{k=1}^{t} a_k.$$

Writing it in this form reminds us that the series' non-stationarity (in this case, variance non-stationarity) is driven by the cumulation of random shocks over time.

Then the differenced series, $Y_t - Y_{t-1}$, becomes

$$\begin{aligned}\Delta Y_t &= [Y_0 + \sum_{k=1}^{t} a_k] - [Y_0 + \sum_{k=1}^{t-1} a_k] \\ &= (a_1 + a_2 + \cdots + a_{t-1} + a_t) - (a_1 + a_2 + \cdots + a_{t-1}) \\ &= a_t\end{aligned}$$

Because all of the past shocks drop out, we are left with only our random error term. By construction, it has an expected value of 0 and a constant variance of σ^2, and so is both mean- and variance-stationary. Differencing has transformed the series into a stationary process.

3.5.2 Unit Root Check 1: Is the Differenced Series Stationary?

It is possible, however, to have higher-order difference non-stationarity, and so it is important to verify that the differenced series is indeed stationary. The differenced DJIA series is shown in Figure 3.9.

It appears to be mean stationary, and the variance of the series does not appear to increase over time as the variance of unit root processes do. To ensure that first-differencing is sufficient to achieve stationarity, we follow the same Elder–Kennedy strategy using the differenced series.

The results from estimating Equation (3.17) using the differenced DJIA series are presented in the second row of Table 3.4. In this case, the test statistic is −15.8, well below the 1% critical value of −3.460, so we are able to reject the null hypothesis of a unit root in the differenced series. Had we not been able to reject the null hypothesis, the next step would be to second-difference the series and perform the same check.

3.5.3 Unit Root Check 2: Could it Have Been a Deterministic Trend Instead?

Given the low power of our unit root tests, a second important check involves examining the autocorrelation function of the differenced series.

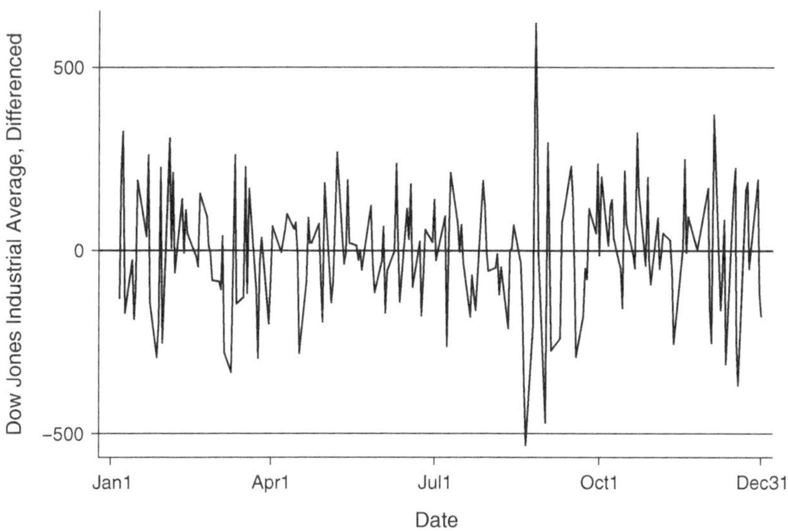

Figure 3.9 The Dow Jones Industrial Stock Index Series Differenced: January 2–December 31, 2015.

While the particulars of this check will not be discussed until Chapter 4, the logic is as follows.

Suppose that the true DGP had a deterministic trend rather than a stochastic trend (unit root). Then it would take the same general form as Equation (3.15),

$$Y_t = \theta_0 + \beta t + a_t$$

Taking the first difference of this series,

$$\begin{aligned} Y_t - Y_{t-1} &= [\theta_0 + \beta t + a_t] - [\theta_0 + \beta(t-1) + a_{t-1}] \\ \Delta Y_t &= \beta + a_t - a_{t-1}. \end{aligned}$$

Notice that in this case, the differenced series includes both the contemporaneous a_t and its lag, a_{t-1}. In Chapter 4, we will see that the added a_{t-1} term makes this a first-order moving average process. The implied coefficient of "1" on a_{t-1} is a telltale sign of "overdifferencing" – differencing an equation that does not have a unit root.

3.6 Extensions of the Dickey–Fuller Test

The assumptions that the error terms $a_1, a_2, \ldots a_t$ in each D–F estimating equation are independent and have a constant variance through the series are often violated in practice. Augmented Dickey–Fuller tests and Phillips–Perron tests, respectively, address these concerns.

3.6.1 Augmented Dickey–Fuller Tests

The basic Dickey–Fuller (D–F) test is appropriate only if the error terms, a_t's in our series, are independent. This is a strong assumption in most cases, since one might guess that at least some of the omitted variables from any model we might construct are autocorrelated. Augmented Dickey–Fuller (ADF) Tests take this into account by including lagged values of the differenced series as independent variables in the estimating equation. In the case of the most general unit root process, this entails estimating Equation (3.24)

$$\Delta Y_t = \alpha Y_{t-1} + \theta_0 + \beta t + \sum_{i-1}^{k} \gamma_i \Delta Y_{t-i} + a_t \tag{3.24}$$

instead of Equation (3.17). The modification is identical across D–F estimating equations, as shown in Table 3.2. Also, the same D–F τ distributions are used for the tests.

The critical question becomes how many lags of the differenced series should be included – put another way, what should k be? If too few lags are included, we will not account for the autocorrelation in the residuals. Additional lags, however, reduce the power of the tests both because the estimation of additional parameters reduces the test's degrees of freedom, and because an observation must be dropped for each period added.

A simple way to select lag length is to start by estimating an ADF model with more parameters than likely needed, and then drop terms as indicated by t-tests. For seasonal data (e.g., data collected monthly or quarterly) the lags should cover at least one season – that is, at least 4 lags for quarterly data and 12 lags for monthly data. Covering 2–3 seasons is preferable. If there are not enough observations to include all lags, all intermediate lags need not be included. For example, with quarterly data, we might include lags 1–4, 8 and 12. For non-seasonal data, 3 lags is typically a reasonable starting point. If the third lag is statistically significant, though, adding lags is in order.

The results of the ADF tests for the DJIA series, starting with three lags and then successively dropping terms, are shown in Table 3.5.

The results of the estimation for $k = 3$ lags – i.e., adding $\gamma_3 \Delta Y_{t-3}$, $\gamma_2 \Delta Y_{t-2}$ and $\gamma_1 \Delta Y_{t-1}$ to the basic D–F estimating equation – are presented in the first two rows. None of the estimated γ coefficients is statistically significant. Because non-seasonal data typically have the strongest autocorrelations at the lowest lags, and additional terms reduce the power of all tests, we will drop the third lag and re-estimate.

The estimates from the equation with $k = 2$ lags do not indicate the additional terms are needed, either; neither γ estimate is statistically significant. Dropping the second lag and reestimating does not change our inference; even with only one lag included, the estimated γ_1 term is not statistically significant. As a final check, we should examine the residuals from our preferred estimating equation (now the ordinary D–F Equation (3.17)). The associated autocorrelation function (described in Chapter 4) should not indicate the presence of autocorrelation in the residuals, and the Q-tests described in Chapter 4 should support this inference.

3.6.2 Phillips–Perron Tests

In addition to assuming that the a_t's in the D–F estimating equations are independent, we also assume that they have constant variance across the series. Phillips–Perron tests are robust to violations of both the independence and the homoskedasticity assumptions. While they are constructed by estimating the same D–F estimating equations – (3.17), (3.19) and (3.20) – and the same D–F τ distributions and/or MacKinnon distribution are used to evaluate the

Table 3.5 Augmented Dickey–Fuller tests for the DJIA series.

N of lags	α	θ_0	β	γ_1	γ_2	γ_3	Test Statistic	1% Critical Value	5% Critical Value	10% Critical Value	MacKinnon P-value
3	−0.052	930	−0.185	0.022	−0.068	−0.018	−2.386	−3.991	−3.430	−3.130	0.3871
	(0.018)	(0.018)	(0.258)	(0.738)	(0.287)	(0.784)					
2	−0.053	967	−0.206	0.021	−0.075		−2.515	−3.990	−3.430	−3.130	0.3206
	(0.013)	(0.012)	(0.201)	(0.740)	(0.239)						
1	−0.057	1029	-−0.206	0.030			−2.725	−3.990	−3.430	−3.130	0.2258
	(0.007)	(0.007)		(0.197)	(0.632)						

P-values associated with ordinary t-tests are reported in parentheses.

test-statistic, the test-statistic incorporates a non-parametric correction to the t-statistic. Because Phillips–Perron tests are less powerful than the ordinary D–F or the ADF tests, ADF tests are preferred unless there is reason to suspect heteroskedasticity in the residuals of the D–F estimating equation. Series with structural breaks – dramatic changes in the parameters of the series – are a common application.

In the case of the DJIA, a stock market crash in August was associated with a precipitous drop in the series, and then a recovery. While the pre-crash series varied between 17 164 and 18 312, on August 20 the series dropped below 17 000 and reached a low of 15 666 before recovering. The volatility of the series for this short stretch is reflected in the variability of the differenced DJIA series that is used in the D–F estimating equation, as seen in Figure 3.9.

Since the same D–F estimating equation, (3.17), is employed, the parameter estimates are identical to those for the D–F test presented in the top row of Table 3.4. The cutpoints for statistical significance are also identical. Only the test statistic and its significance level will be different. The estimated Phillips–Perron test statistic is −2.504, while the ordinary D–F test statistic was −2.631. Neither allows us to reject the null hypothesis of a unit root.

3.7 Seasonal Non-stationarity

Seasonal variation in the level of the series can also be a source of mean non-stationarity. Diagnosing seasonal trend (as opposed to seasonal components that do not result in a trend) is most easily done in conjunction with the autocorrelation function presented in Chapter 4 and estimation techniques presented in Chapter 5, and will be examined in detail in those chapters. Two points can be made briefly, though. First, series with seasonal components are easily recognized in time series plots. Consider Figure 3.10, showing monthly averages of the atmospheric concentration of carbon dioxide (CO_2) in parts per million. CO_2 concentrations are strongly affected by plants' photosynthesis. Consequently, CO_2 concentrations increase during the North's winter months when many trees have lost their leaves, and decline during the North's Spring and Summer months when foliage absorbs CO_2. This pattern is visually apparent in the time series plot.

There is also a clear trend in the plot. If the trend is driven by the seasonal component, seasonal differencing rather than first-differencing is in order; for example, in this case we would examine whether the series $\Delta_S^{12} Y_t = Y_t - Y_{t-12}$ is a better model than $\Delta Y_t = Y_t - Y_{t-1}$.

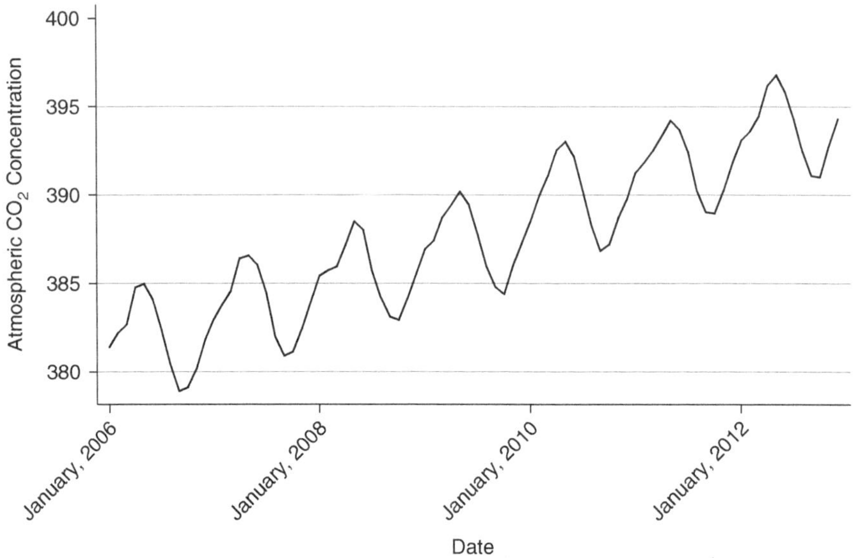

Figure 3.10 Average atmospheric CO_2 concentrations, parts per million by volume, 2006–2012. (Source: Dr. Pieter Tans, NOAA/GML (www.esrl.noaa.gov/gmd/ccgg/trends/) and Dr. Ralph Keeling, Scripps Institution of Oceanography (scrippsco2.ucsd.edu/).

3.8 Variance Non-stationarity

Thus far we have restricted our attention to mean non-stationarity – i.e., we have examined methods for testing whether or not the level of the series increases or decreases over time. Because differencing a unit root process addresses both mean non-stationarity and variance non-stationarity, we commonly solve both problems at once by correctly diagnosing and addressing the unit root. In the case of the Boston Robbery series, though, we rejected the null hypothesis of a unit root and inferred that the series has a deterministic trend and variance non-stationarity. Since modeling the deterministic trend does not, by itself, address the variance non-stationarity, further investigation is required.

3.8.1 Count Data

A second common source of variance non-stationarity in time series data is the use of count data. Because count data are bounded below, the variance of the series when they are close to zero is necessarily lower than the variance of the series closer to their maximum values.[15] The Boston Robbery series is a typical example – it is a count of the number of armed robberies in each month, and visual inspection of Figure 3.5 indicates a low spread of the values in the early months when the counts were small numbers (less than 100), and a much larger spread late in the series when the values exceeded 300.

Taking the natural logarithm of count data addresses the variance non-stationarity due to the count. A plot of the logged Boston Robbery series is shown in Figure 3.11.

While the series still trends upward, logging the series has eliminated the obvious variance non-stationarity. Does this change the inference we drew regarding the unit root? Re-estimating the D–F Equation (3.17) using the pre-intervention logged series, we find

$$\Delta \ln(Y_t) = \underset{(-5.51)}{-0.44} \ln(Y_{t-1}) + \underset{(5.21)}{1.58} t + \underset{(-0.64)}{-4.58} + a_t \tag{3.25}$$

where the numbers in parentheses are the estimated t-statistics. Because the test statistic −5.51 is well below the cutpoint for the 1% critical value, −4.037, we still reject the null hypothesis and will model a deterministic trend.

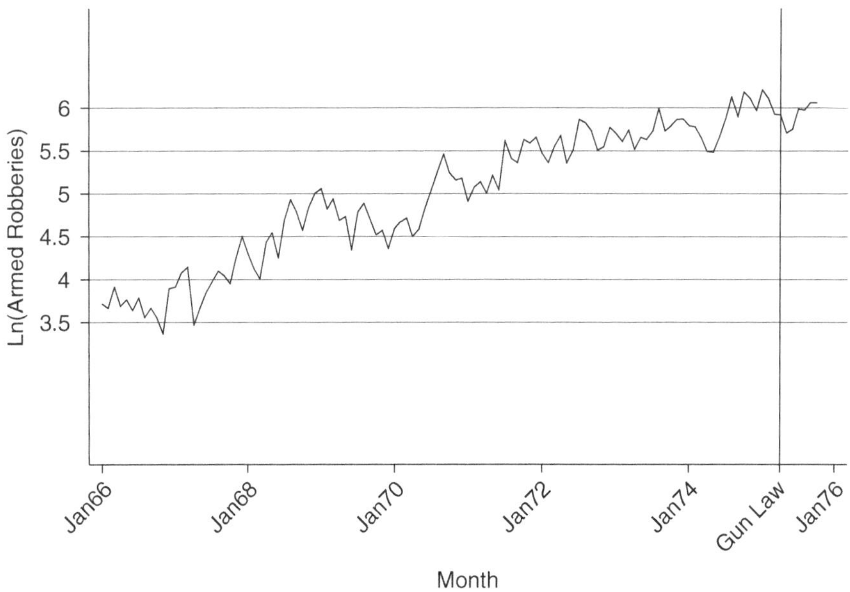

Figure 3.11 Boston Robberies, Logged, January 1966 – October 1975.

15 Readers familiar with the Poisson distribution that is often used to model count data may remember that it is characterized by a variance proportional to its mean.

3.8.2 Aside on Autoregressive Conditional Heteroskedasticity Models

A prominent feature of the DJIA series was the dramatic, unexpected decline in the series in mid-August and slow return to its previous level. In addition to the temporary level change, the variance of the series was much higher during this episode. Technically this is not a violation of variance non-stationarity because the change in variance was not expected; i.e., Equation (3.3) still holds. This pattern, though – short periods of high volatility interspersed with longer stretches of relative stability – is not uncommon in financial time series. Since anticipating risk entails predicting not only the change in the level of the series, but how wide the swings may be, they are an important consideration for market analysts. Autoregressive Conditional Heteroskedasticity (ARCH) models use the error terms (a_t's) in the previous periods to predict the variability, and in the case of the ARCH-M model, to help predict the level of the series as well.

3.9 Summary

- Properly specified time series models account for the time series properties of the series. Mean-stationarity and variance-stationarity are discussed in this chapter.
- Trends in observed time series (mean non-stationarity) may be either deterministic or stochastic. In the same way that it's possible to flip a disproportionate number of heads in a series of coin flips, it is possible for random shocks to be disproportionately positive (or negative). Their cumulation over time can cause apparent trends in a series even when its underlying DGP is mean-stationary.
- DGP's with stochastic trends are one type of unit root process. Unit root tests are a means of evaluating whether or not there is a stochastic trend.
- Dickey–Fuller tests and their cousins, augmented Dickey–Fuller tests and Phillips–Perron tests, are commonly-used tools to discern whether an apparent trend is deterministic or stochastic.
- The tests have low power, and share the unusual feature that the null hypothesis is the presence of a unit root. The Elder–Kennedy algorithm for employing and evaluating the tests is introduced as a means of mitigating the danger of incorrect inferences.
- Seasonal mean non-stationarity will be addressed in Chapter 4.
- The unit root processes that cause stochastic trends also cause variance non-stationarity, so transforming the unit root process by differencing the series typically addresses both the mean and the variance non-stationarity.
- Count data are another common source of variance non-stationarity. The log transformation eliminates this form of variance non-stationarity.

References

Dickey D.A. (1984). Powers of Unit Root Tests. *Proceedings of Business and Economic Statistics Section, American Statistical Association*. 489–493.

Dickey D.A. and Fuller W.A. (1979). Distribution of the Estimates for Autoregressive Time Series with a Unit Root. *Journal of the American Statistical Association* 74(June):427–431.

Dickey D.A. and Fuller W.A. (1981). Likelihood Ratio Statistics for Autoregressive Time Series with a Unit Root. *Econometrica* 49(4):1057–1072.

Elder J. and Kennedy P.E. (2001). F Versus t Tests for Unit roots. *Economics Bulletin* 3:1–6.

Elder J. and Kennedy P.E. (2001). Testing for Unit Roots: What Should Students Be Taught? *The Journal of Economic Education* 32(2):137–146.

Holden D. and R. Perman (1994). Unit roots and cointegration for the economist. In *Cointegration for the Applied Economist* (ed. B.B. Rao), 47–112. New York: St. Martin's.

McCleary R. and Hay R.A. (1980). Applied Time Series for the Social Sciences. Beverly Hills: Sage Publications.

Perron P. (1988). Trends and random walks in macroeconomic time series. *Journal of Economic Dynamics and Control* 12(12):297–232.

Phillips P. and Perron P. (1988). Testing for a Unit Root in Time Series Regression. *Bimetrika* 75(June):335–346.

Tans P. and Keeling R. Carbon Dioxide Research Group, Scripps Institution of Oceanography, NOAA Global Monitoring Laboratory. (2014). Trends in Atmospheric Carbon Dioxide. Downloaded from https://www.esrl.noaa.gov/gmd/ccgg/trends/ March 3, 2014.

United States Renal Data System 2018 USRDS annual data report: Epidemiology of kidney disease in the United States. National Institutes of Health, National Institute of Diabetes and Digestive and Kidney Diseases. Bethesda, MD. Data taken from Vol 2, Chapter 2, downloaded from https://usrds.org/2018/view/v2_02.aspx Nov. 11, 2019.

Yule G.U. (1926). Why do we sometimes get Nonsense-Correlations between Time-Series? – A Study in Sampling and the Nature of Time-Series. *Journal of the Royal Statistical Society* 89(1):1–63.

4

Properties of Time Series: Autocorrelation

4.1 Rethinking Autocorrelation

In introductory regression courses, autocorrelation is introduced as a potential problem afflicting the error term in a regression model such as this

$$Y_t = \alpha + \beta X_t + \epsilon_t \tag{4.1}$$

and our attention is frequently restricted to first-order autocorrelation, i.e., correlation between ϵ_t and ϵ_{t-1}.

It is, however, more useful to recognize that in typical time series applications, the dependent variable Y_t is autocorrelated – e.g., Y_t is correlated with Y_{t-1}. Consider a stock price or voters' approval rating of a politician. In the absence of a dramatic event, we would be surprised if there were a dramatic change from one day to the next – that is, we expect today's value, Y_t, to be relatively close to yesterday's value, Y_{t-1}. We may not be sure of the precise reasons for this – for example, whether voters simply maintain fixed assessments of politicians until shaken by a random event, or if the assessments tend to be stable in the short run because they are driven by a number of other factors that are stable in the short run (e.g., economic performance, public statements, and/or news about policy choices). Either way, though, autocorrelation is almost always a central feature of the dependent variables we are modeling when working with time series.

Consequently, every model using time series data is predicated upon critical choices, conscious or not, about how this autocorrelation in Y_t is modeled. In Equation (4.1), autocorrelation in Y_t that is not accounted for by autocorrelation in X_t will appear in the error term, ϵ_t. Remember, though, that

Applied Time Series Analysis for the Social Sciences: Specification, Estimation, and Inference, First Edition. Regina M. Baker.

Companion website: www.wiley.com/go/baker

autocorrelation is often[1] driven by omitted variables. Relegating the autocorrelation to the error term, then, exposes us to the risk of confounding – that our coefficient estimates will be plagued by omitted variables bias because the source of the autocorrelation addressed in the error term is correlated with an included variable.

Different approaches address this problem and the problem of spurious correlations in different ways. In classical time series analysis, also known as ARIMA or Box-Jenkins modeling, a preliminary step involves creating theoretical models of the autocorrelation of both the dependent variable and any independent variables and controlling for their shared structure. More recent approaches – including the general-to-specific modeling, LSE approach, and VAR models covered in chapters 9-13 – use lags of the dependent and independent variables in the estimating equations to ensure that autocorrelation is covered by included variables rather than relegated to the error term.[2]

In all of these approaches, diagnosing autocorrelation plays two roles. First, after transforming the series to address any non-stationarity, the autocorrelation structure of the transformed series is examined to make preliminary guesses about the DGP. Second, the autocorrelation structure of the estimated error terms from each candidate model is examined to evaluate the model. The tools used in both cases include the sample autocorrelation function (SACF) and sample partial autocorrelation function (SPACF), introduced later in the chapter. These are compared to autocorrelation functions (ACFs) and partial autocorrelation functions (PACFs) from archetypal DGPs.

An illustration may help introduce the process. Remember that in Chapter 3, we were concerned that because the unit root test for the DJIA series just missed statistical significance, our inference that the series required differencing may be incorrect. In section 3.5.3, we saw that if we erred by differencing the series when it had a deterministic trend instead of a stochastic trend, the error term of the differenced series would exhibit a particular form of autocorrelation:

$$\Delta DJIA_t = \beta + a_t - a_{t-1} \tag{4.2}$$

where β was the coefficient on the deterministic trend term and a_t was our white noise process. As part of evaluating the possibility of overdifferencing, we will compare the graphic representation of the SACF of $\Delta DJIA$, shown in

1 Always? At its root, this is a philosophical question about whether any event is truly random, or if some of them are merely beyond our understanding.

2 This, of course, changes the interpretation of the estimates – a topic addressed in Chapters 9–13.

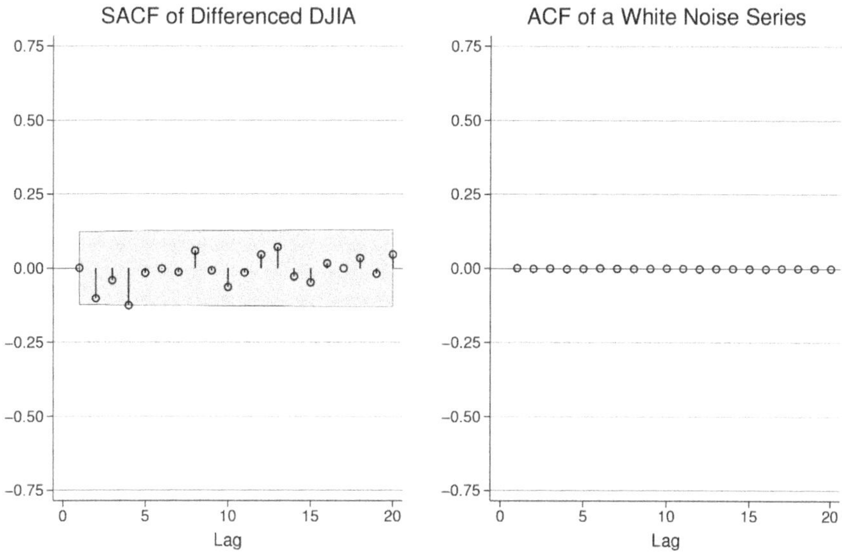

Figure 4.1 Example of a SACF and an ACF Shaded area indicates 95% confidence bands.

the left panel of Figure 4.1 with the ACF of a white noise process, depicted in the right panel of Figure 4.1.

We will examine this case in detail later in the chapter. To set the stage for this type of comparison and for the use of SACFs to make preliminary guesses about the DGP, the chapter will proceed as follows. First, Wold's Theorem will be introduced as the justification for modeling any time series as a moving average (MA) process, and MA processes will be introduced in more detail. Next, ACFs and SACFs are defined formally, and ACFs of white noise and MA processes are derived. Later in the chapter, autoregressive (AR) processes are introduced as an alternative to modeling series in MA terms, and their ACFs are derived as well. A second tool, the partial autocorrelation function (PACF), is introduced to help distinguish between DGPs whose ACFs are ambiguous. The chapter closes with discussions of seasonality and mixed processes.

4.2 Modeling Autocorrelation: Wold's Theorem

Modeling a time series in terms of MA and AR components is not merely an attempt to approximate the series; Wold's Theorem provides the justification for modeling any time series as the weighted sum of a series of random shocks.

Wold's Theorem: Every weakly stationary, purely non-deterministic stochastic process $(Y_t - \theta_0)$ can be written as a linear combination of a sequence of uncorrelated random variables:

$$Y_t - \theta_0 = a_t + \theta_1 a_{t-1} + \theta_2 a_{t-2} + \ldots + \theta_k a_{t-q} \tag{4.3}$$

or alternatively,

$$Y_t - \theta_0 = \sum_{j=0}^{\infty} \theta_j a_{t-j}, \quad \theta_0 = 1,$$

where

- Y_t is the series we're studying,
- θ_0 is the series mean,
- The a_{t-j}'s are random shocks (sometimes referred to as "white noise,") $a_{t-j} \sim N(0, \sigma^2)$, and
- The θ_j's (theta's) are coefficients. Because Wold referred to them as ψ's rather than θ's, they are sometimes referred to as psi-weights.

Remember that although we have written Y_t here – common notation for a dependent variable – the characterization of autocorrelation that we will examine in this chapter is applicable to any time series – that is,it is equally applicable to explanatory variables and the error terms from estimated equations. Also, notice that it doesn't make use of any substantive knowledge of the problem at hand. Suppose, for example, that our series of interest is the percentage of Americans each month who approve of the president's job performance. When we model the series as in Equation (4.3), we are saying nothing more than "the percentage of Americans who approve of the president's job performance this month can be modeled as an average approval rating that is shifted by random occurrences in the current and preceding months." The ψ-weights indicate the relative impact of the timing of the random events.

4.3 Moving Average Processes

Wold's representation in Equation (4.3) is the general form of a "Moving Average (MA) Process," although MA processes, by convention, are written with their means on the right-hand-side of the equation and with the MA components subtracted from the mean:

$$Y_t = \theta_0 + a_t - \theta_1 a_{t-1} - \theta_2 a_{t-2} - \cdots - \theta_q a_{t-q}$$

or

$$Y_t = \theta_0 + (1 - \theta_1 L - \theta_2 L^2 - \ldots - \theta_q L^q) a_t$$

MA processes of order q use lags of the error term as high as the order q to predict the current value of the series:

The order q is defined by the highest order lag present in the model, not by the number of lags present.

4.3.1 The MA(1) Process

The simplest form, an MA(1) process, is represented as follows:

$$Y_t = \theta_0 + a_t - \theta_1 a_{t-1}$$

or

$$Y_t = \theta_0 + (1 - \theta_1 L) a_t.$$

If our sample autocorrelation function indicated that our presidential approval series could be well-represented by an MA(1) process, we might infer that only random events in the previous month shifts presidential approval; random events from just two months earlier have no impact, as the associated θ-coefficients (or ψ-weights) are implicitly equal to 0.

We call this an MA(1) series because the highest order lag of *at* is the first-order lag, a_{t-1}. It is important to remember that Wold's Theorem applies only to stationary series. The term "Bounds of Invertibility" is used to refer to the restrictions on the θ terms that must apply for an MA process to be stationary. In the case of an MA(1) series, the Bounds of Invertibility are limited to

$$-1 < \theta_1 < +1.$$

4.3.2 The MA(2) Process

MA(2) processes include a lag of order 2, but no higher. The general form is

$$Y_t = \theta_0 + a_t - \theta_1 a_{t-1} - \theta_2 a_{t-2},$$

or

$$Y_t = \theta_0 + (1 - \theta_1 L - \theta_2 L^2) a_t.$$

The equation need not include the a_{t-1} term to be an MA(2) process, though. The model

$$Y_t = \theta_0 - \theta_2 a_{t-2}$$

is also an MA(2) process.

The bounds of invertibility for an MA(2) process are as follows:

$$-1 < \theta_2 < +1,$$

$$\theta_1 + \theta_2 < +1, \text{ and}$$

$$\theta_1 - \theta_2 < +1.$$

An MA(2) process will be non-stationary if any one of these conditions is not met.

Although Wold's Theorem tells us that any series can be modeled as an MA process, in practice, the need to work with MA orders higher than two is rare because of an important relationship between MA processes and autoregressive (AR) processes, described later in the chapter. Before turning to AR processes, though, let's introduce the autocorrelation function and show how it can be used to decide whether a particular series is better represented by a white noise process, an MA(1) process, an MA(2) process, or a higher-order MA process.

4.4 The Autocorrelation Function (ACF) and Sample Autocorrelation Function (SACF)

4.4.1 Definitions

The ACF of a stationary DGP Y_t is the set of correlations, ρ, between Y_t and lags of itself. In other words, the ACF to lag k includes the following elements:

$$\begin{aligned} \rho_1 &= \text{corr}(Y_t, Y_{t-1}) \\ \rho_2 &= \text{corr}(Y_t, Y_{t-2}) \\ \rho_3 &= \text{corr}(Y_t, Y_{t-3}) \\ &\vdots \\ \rho_k &= \text{corr}(Y_t, Y_{t-k}) \end{aligned}$$

The right panel of Figure 4.1 is a graphical representation, or correlogram, of the ACF to lag 20 of a white noise process. We will show below why, for this series,

$$\rho_1 = \rho_2 = \cdots = \rho_k = 0$$

for all k. To simplify derivations, we will make heavy use of the ACF formula written in terms of covariances:[3]

$$\rho_k = \frac{\text{Cov}(Y_t, Y_{t-k})}{\text{Var}(Y_t)} \tag{4.4}$$

3 A review of expectation operators, covariance and correlation is provided in Appendix B.

Given a stationary series *Yt* with mean θ_0 and constant variance σ^2, the autocovariance at lag k is defined as

$$\gamma_k = \text{Cov}(Y_t, Y_{t-k}) = E(Y_t - \theta_0)(Y_{t-k} - \theta_0). \quad (4.5)$$

Before deriving the ACF for the white noise process, notice that the SACF must be calculated slightly differently. While our DGP extends infinitely into both the past and future, the data we have for any particular realization are finite, and so we lose *k* observations when we calculate a sample autocorrelation, *r*, at lag *k*.

$$r_k = \frac{\Sigma_{t=k+1}^{N}(Y_t - \bar{Y})(Y_{t-k} - \bar{Y})}{\Sigma_{t=1}^{N}(Y_t - \bar{Y})^2} \quad (4.6)$$

$$= \frac{\hat{\text{Cov}}(Y_t, Y_{t-k})}{\hat{\text{Var}}(Y_t)} \quad (4.7)$$

The SACF for the differenced DJIA series is presented in the left panel of Figure 4.1. The shaded band represents the 95% confidence interval for each of the autocorrelations. Under the null hypothesis that $\rho_k = 0$, the variance of the sample autocorrelations is calculated as

$$\frac{1}{n}(1 + 2r_1^2 + 2r_2^2 + \cdots + 2r_j), \quad j < k$$

and confidence bands are constructed using this variance.

The SACF presented in Figure 4.1 includes values for 20 lags. Under the null hypothesis that the DGP is a white noise process, we would expect one of these 20 to be statistically significant at the 95% level just by chance. Consequently, some judgment is required when interpreting SACFs: we investigate spikes in "sensible" places and tend to dismiss the others when their number can be explained by chance. "Sensible" places include the first few lags, lags corresponding to any seasonal component, and any lags explicable by the substance of the problem.

Additionally, Q-tests provide joint tests of significance for lags up to a given number.

4.4.2 Q-tests for SACFs

The Box-Ljung Q-statistic is calculated as follows:

$$Q = T(T+2)\sum_{j=1}^{k} r_j^2 / (T - j)$$

where

- T is the number of time points,
- j indexes the lag, and
- k is the maximum lag used in the ACF.

The Q-statistic follows a χ^2 distribution with k degrees of freedom. If the calculated Q exceeds the critical value of χ^2 with k degrees of freedom, we infer that at least one of the autocorrelations up to and including that lag is statistically significant. Statistical software packages that include Q-statistics as part of their ACF output typically include Q-tests for several lags. The sample autocorrelation, r, and associated Q-tests through lag 20 for the differenced DJIA series are presented in Table 4.1.

r_1, the estimated autocorrelation at lag 1, is 0.001. The associated Q-test tests the null hypothesis that ρ_1, the autocorrelation at lag 1 from the DGP, is equal to 0. The calculated Q-statistic was less than 0.001, and not statistically significant ($p = .99$); we cannot reject the null hypothesis. r_2, the estimated autocorrelation at lag 2, is −0.102. In this case, the associated Q-test tests the null hypothesis that *both* ρ_1 and ρ_2 are equal to 0. We are unable to reject the null hypothesis; the Q-statistic is 2.66, $p = 0.26$.

For this series, only r_4 appeared to reach the confidence limit. However, we expect one in twenty autocorrelations to be statistically significant by chance, the lower lags are not statistically significant, and the fourth lag does not correspond to a seasonal marker; we have little reason to accord it any credence. The Q-test at lag 4 provides additional support: we are unable to reject the null hypothesis that the first four autocorrelations, $\rho_1 - \rho_4$, are all equal to 0, $p = 0.13$.

Table 4.1 Autocorrelations and Q-tests for the Differenced DJIA Series.

Lag	r	Q	P-value	Lag	r	Q	P-value
1	0.001	<0.001	0.99	11	−0.014	9.32	0.59
2	−0.102	2.66	0.26	12	0.046	9.89	0.63
3	−0.040	3.08	0.38	13	0.072	11.27	0.59
4	−0.126	7.18	0.13	14	−0.028	11.47	0.65
5	−0.016	7.24	0.20	15	−0.048	12.07	0.67
6	−0.001	7.25	0.30	16	0.017	12.15	0.73
7	−0.013	7.29	0.40	17	0.000	12.15	0.79
8	0.059	8.20	0.41	18	0.034	12.47	0.82
9	−0.007	8.22	0.51	19	−0.018	12.56	0.86
10	−0.063	9.27	0.51	20	0.047	13.16	0.87

4.4.3 ACF for a White Noise Process

This exercise – comparing an SACF to the ACF for a white noise process – is ubiquitous in time series work because it is the standard diagnostic for testing whether the autocorrelation in a dependent variable is "covered by" the included independent variables; remember that if the autocorrelation is left in the error term, we risk omitted variables bias in our coefficients.

Let's take a closer look at the white noise process and derive its ACF. A white noise process Y_t take the form

$$Y_t = \theta_0 + a_t$$

where θ_0 is the mean of the series, $a_t \sim \text{iid}(0, \sigma^2)$, and "iid" is an abbreviation for "independently and identically distributed." In the case of an estimated error term, $\theta_0 = 0$.

We'll begin by deriving the autocovariances γ_k using Equation (4.5), and then substitute them into Equation (4.4) for each autocorrelations, ρ_k.

First, notice that $\gamma_0 = \text{Var}(Y_t) = \sigma^2$.

$$\begin{aligned}\gamma_0 &= E[(Y_t - \theta_0)(Y_t - \theta_0)] \\ &= E(\theta_0 + a_t - \theta_0)(\theta_0 + a_t - \theta_0) \\ &= E(a_t^2) \\ &= E[(a_t - 0)(a_t - 0)] \\ &= Var(a_t) \\ &= \sigma^2\end{aligned}$$

You will use the fact that since the mean of a_t is zero, $Var(a_t) = E(a_t^2) = \sigma^2$ in many derivations.

$$\begin{aligned}\gamma_1 &= E(Y_t Y_{t-1}) \\ &= E[(\theta_0 + a_t - \theta_0)(\theta_0 + a_{t-1} - \theta_0)] \\ &= E[a_t a_{t-1}] \\ &= 0\end{aligned}$$

By construction, a_t and a_{t-k} are independent for all k. Then

$$E(a_t a_{t-1}) = E(a_t a_{t-2}) = \cdots = E(a_t a_{t-k}) = 0$$

for all $k \neq 0$.

Then the autocorrelations

$$\begin{aligned}\rho_k &= \frac{\gamma_k}{\gamma_0} \\ &= \begin{cases} 1 \text{ if } k=0 \\ 0 \text{ if } k>0 \end{cases}\end{aligned} \tag{4.8}$$

That is, autocorrelations at all lags > 0 are equal to 0.

4.5 ACFs for MA Processes

MA processes also have easily interpretable ACFs.

4.5.1 The MA(1) Process

Recall that an MA(1) process takes the form

$$Y_t = \theta_0 + a_t - \theta_1 a_{t-1}$$

Once again, we will begin by deriving the autocovariances γ_k for the first few lags using Equation (4.5), and then substitute them into Equation (4.4) for each of the first few autocorrelations, ρ_k. If you look carefully at the derivations of the ACF for the white noise series, you'll notice that the mean of the series quickly drops from the calculations. For simplicity's sake, and with no loss of generality, we will work with a zero-mean series, $Y_t = a_t - \theta_1 a_{t-1}$.

$$\begin{aligned}
\gamma_0 &= E(Y_t Y_t) \\
&= E[(a_t - \theta_1 a_{t-1})(a_t - \theta_1 a_{t-1})] \\
&= E[a_t^2 - 2\theta_1 a_t a_{t-1} + \theta_1^2 a_{t-1}^2] \\
&= E[a_t^2] - 2\theta_1 E[a_t a_{t-1}] + \theta_1^2 E[a_{t-1}^2] \\
&= \sigma^2(1 + \theta_1^2)
\end{aligned}$$

$$\begin{aligned}
\gamma_1 &= E(Y_t Y_{t-1}) \\
&= E[(a_t - \theta_1 a_{t-1})(a_{t-1} - \theta_1 a_{t-2})] \\
&= E[a_t a_{t-1} - \theta_1 a_t a_{t-2} - \theta_1 a_{t-1}^2 + \theta_1^2 a_{t-1} a_{t-2}] \\
&= -\theta_1 E[a_{t-1}^2] \\
&= -\theta_1 \sigma^2
\end{aligned}$$

$$\begin{aligned}
\gamma_2 &= E(Y_t Y_{t-2}) \\
&= E[(a_t - \theta_1 a_{t-1})(a_{t-2} - \theta_1 a_{t-3})] \\
&= E[a_t a_{t-2} - \theta_1 a_t a_{t-3} - \theta_1 a_{t-1} a_{t-2} + \theta_1^2 a_{t-1} a_{t-3}] \\
&= 0
\end{aligned}$$

Substituting these into Equation (4.4) and simplifying,

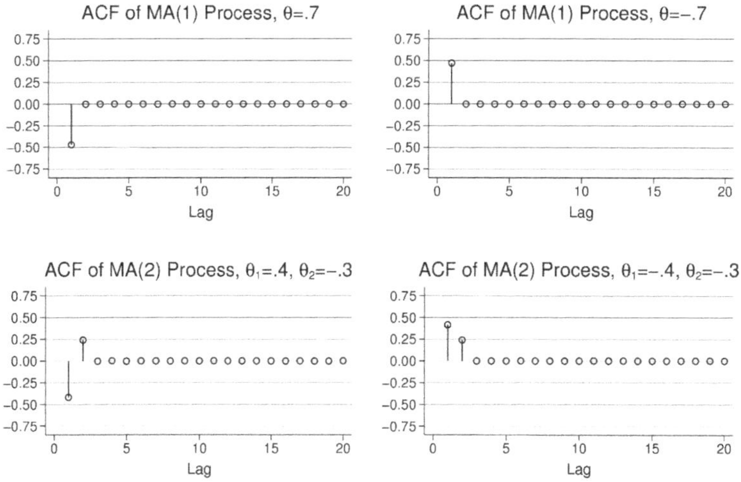

Figure 4.2 ACFs for Four MA Series.

$$\rho(1) = \frac{\gamma_1}{\gamma_0} = -\frac{\theta_1}{1+\theta_1^2}$$

$$\rho(2) = \frac{\gamma_2}{\gamma_0} = 0$$

$$\vdots$$

$$\rho(k) = \frac{\gamma_k}{\gamma_0} = 0$$

The ACF for an MA(1) process has a spike at lag 1, but is 0 for all lags greater than one. Examples are provided in the top row of Figure 4.2.

4.5.2 The General Form for the ACF of an MA(q) Process

In the exercises, you will be asked to derive the autocovariances and autocorrelations for an MA(2) process using the same method. When you remember that all terms involving expected values of cross-products – e.g., $E[a_t a_{t-1}]$ – will equal 0 and drop out, it's fairly straightforward to derive a general form. For any MA(q) process,

$$\gamma_0 = (1+\theta_1^2+\theta_2^2+\cdots+\theta_q^2)\sigma^2$$

$$\gamma_k = (-\theta_k+\theta_1\theta_{k+1}+\cdots+\theta_{q-k}\theta_q)\sigma^2$$

where $k = 1,2,\ldots q$.

Then the general form for the autocorrelations, ρ_k of an MA(q) process is

$$\rho(k) = \frac{\gamma_k}{\gamma_0} = \frac{-\theta_k + \theta_1\theta_{k+1} + \cdots + \theta_{q-k}\theta_q}{1 + \theta_1^2 + \theta_2^2 + \cdots + \theta_q^2}$$

For an MA(2) process –

$$Y_t = a_t - \theta_1 a_{t-1} - \theta_2 a_{t-2}$$

– this becomes

$$\begin{aligned}
\gamma_0 &= (1 + \theta_1^2 + \theta_2^2)\sigma^2 \\
\gamma_1 &= (-\theta_1 + \theta_1\theta_2)\sigma^2 \\
\gamma_2 &= -\theta_2\sigma^2 \\
\gamma_3 &= 0 \\
&\vdots \\
\gamma_k &= 0, \ k > 2
\end{aligned}$$

$$\rho(0) = \frac{(1 + \theta_1^2 + \theta_2^2)\sigma^2}{(1 + \theta_1^2 + \theta_2^2)\sigma^2} = 1$$

$$\rho(1) = \frac{-(\theta_1 + \theta_1\theta_2)\sigma^2}{(1 + \theta_1^2 + \theta_2^2)\sigma^2} = \frac{-\theta_1 + \theta_1\theta_2}{1 + \theta_1^2 + \theta_2^2}$$

$$\rho(2) = \frac{-\theta_2\sigma^2}{(1 + \theta_1^2 + \theta_2^2)\sigma^2} = \frac{-\theta_2}{1 + \theta_1^2 + \theta_2^2}$$

$$\begin{aligned}
\rho(3) &= 0 \\
&\vdots \\
\rho(4) &= 0
\end{aligned}$$

For MA(2) processes, then, the ACF will have spikes at lags 1 and 2, and $\rho_k = 0$ for all higher lags. Illustrations are provided in the bottom row of Figure 4.2.

4.5.3 Example 1: Test Yourself

To test your understanding so far, examine the SACFs in Figure 4.3. For each of the two, decide whether you think the corresponding DGP is an MA(1) process, an MA(2) process or a white noise process. For any suspected MA(1) process, do you think the MA(1) coefficient is positive or negative? You may assume that there are no seasonal effects. The answers follow Example 2.

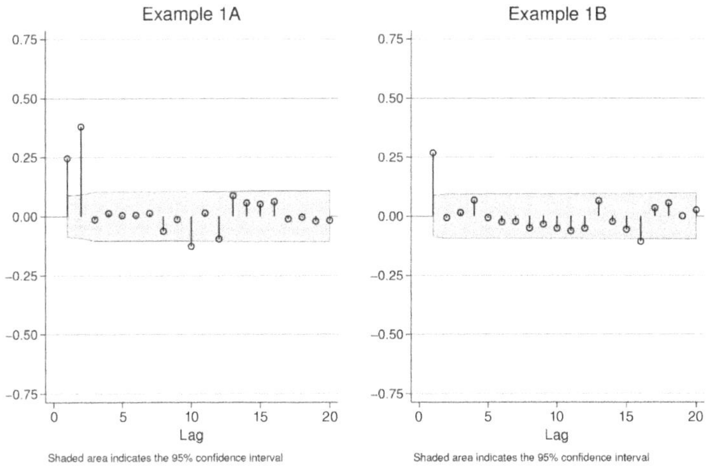

Figure 4.3 SACFs for Example 1 Mystery Series.

4.5.4 Example 2: Did we Overdifference the DJIA Series?

In Chapter 3, section 3.5.3, we saw that if we differenced a series that did not have a unit root, the series would exhibit a particular form of autocorrelation:

$$\Delta DJIA = \beta + a_t - a_{t-1}$$

where β is the coefficient on a deterministic time trend ($\beta = 0$ if there is no deterministic trend) and a_t is our white noise error term. Notice that this is an MA(1) process with an implied $\theta_1 = 1$, which would violate the bounds of invertibility. In practice, the estimate of the MA(1) term is typically slightly less than one, and the overdifferenced series has an ACF with a single spike approximately equal to −0.5 at the first lag, and no further spikes. In Figure 4.1, though, the only spike that comes close to reaching statistical significance, is at lag 4. This SACF bears no resemblance to an MA(1) process. Indeed, the SACF for the differenced DJIA series looks very much like the ACF for a white noise series. We expect one in twenty spikes, on average, to be statistically significant even if the underlying process is white noise, so the spike at lag 4 is unconcerning, and the Q-tests presented in Table 4.1 do not allow us to reject the null hypothesis that for the DGP, all four autocorrelations through lag 4 are 0. We can safely rule out the possibility that we overdifferenced the series.

4.5.5 Example 1: Answers

The SACF in Example 1A has clear spikes at lags 1 and 2, followed by estimated autocorrelations inside the 95% confidence band with only one exception, at lag 10. Since we can assume there is no seasonal effect at lag 10, and we have no substantive explanation, we are inclined to explain it as chance variation. Our best initial guess is that the SACF is from an MA(2) series.

The SACF in Example 1B has one clear spike at lag 1, and only one other lag that exceeds the 95% confidence band – at lag 16. Our best initial guess is that the SACF is from an MA(1) series. Recall that for an MA(1) series $Y_t = \theta_0 + a_t - \theta_1 a_{t-1}$, the autocorrelation at lag 1 is

$$\rho_1 = \frac{-\theta_1}{1+\theta_1^2}$$

Since the spike at lag 1 in Example 1B is positive, the SACF indicates that θ_1 is negative. If you are ambitious and nostalgic for high school algebra, notice that with an estimate of ρ_1, you could use the quadratic formula to derive an associated estimate for θ_1 after rewriting the formula for ρ_1 as follows:

$$\begin{aligned} \rho_1 &= \frac{-\theta_1}{1+\theta_1^2} \\ \rho_1(1+\theta_1^2) &= -\theta_1 \\ \rho_1\theta_1^2 + \theta_1 + \rho_1 &= 0 \end{aligned}$$

4.6 Autoregressive Processes

Although Wold's Theorem does not mention them, autoregressive (AR) processes have a special relationship with MA processes which makes them an alternative to modeling any stationary stochastic series with only MA components. Let's begin with definitions, and then explain the relationship.

Autoregressive processes of order p use p lagged observations of the series to predict the current observation. They take the general form

$$Y_t = \theta_0 + \phi_1 Y_{t-1} + \phi_2 Y_{t-2} + \ldots + \phi_p Y_{t-p} + a_t$$

or

$$(1 - \phi_1 L - \phi_2 L^2 - \ldots - \phi_p L^p) Y_t = \theta_0 + a_t.$$

4.6.1 The AR(1) Process

An AR(1) process, then, can be written

$$Y_t = \theta_0 + \phi_1 Y_{t-1} + a_t$$

or

$$(1 - \phi_1 L) Y_t = \theta_0 + a_t.$$

Like MA processes, AR processes are defined only for stationary series. The restrictions on the ϕ coefficients required for the series to be stationary are referred to as the "Bounds of Stationarity." In the case of AR(1) models, the Bounds of Stationarity is just

$$-1 < \phi_1 < 1$$

exactly what you might expect after studying unit root processes.

4.6.2 AR(2) Processes

The general form of AR(2) processes can be written as

$$Y_t = \theta_0 + \phi_1 Y_{t-1} + \phi_2 Y_{t-2} + a_t$$

or

$$(1 - \phi_1 L - \phi_2 L^2) Y_t = \theta_0 + a_t.$$

Remember that the order of the process (p) is given by the highest order lag, not the number of lags in the series.

$$Y_t = \theta_0 + \phi_2 Y_{t-2} + a_t$$

is an AR(2) process.

The Bounds of Stationarity for an AR(2) process are as follows:

$$-1 < \phi_2 < +1$$

$$\phi_1 + \phi_2 < +1$$

$$\phi_1 - \phi_2 < +1.$$

As in the case of the Bounds of Invertibility for MA(2) processes, all of these restrictions must be met for the series to be stationary.

4.6.3 Moving Average vs. Autoregressive Representations

While Wold's Theorem refers explicitly to MA processes, the relationship between MA and AR processes permits us to represent series more parsimoniously by using a combination of AR and MA terms. To get some intuition behind the reason for this, watch what happens when we re-write an AR(1) series in MA form

$$\begin{aligned} Y_t &= \theta_0 + \phi_1 Y_{t-1} + a_t \\ (1 - \phi_1 L) Y_t &= \theta_0 + a_t \\ Y_t &= \frac{\theta_0}{1 - \phi_1} + \frac{a_t}{1 - \phi_1 L} \end{aligned}$$

Using our lag operator algebra for infinite series, presented in Chapter 2, we can rewrite this as

$$\begin{aligned} Y_t &= \frac{\theta_0}{1-\phi_1} + (1 + \phi_1 L + \phi_1^2 L^2 + \phi_1^3 L^3 + \cdots)a_t \\ &= \frac{\theta_0}{1-\phi_1} + a_t + \phi_1 a_{t-1} + \phi_2 a_{t-2} + \phi_3 a_{t-3} + \cdots \end{aligned}$$

In other words, an AR(1) process is mathematically equivalent to an infinite MA process. The converse is also true: any MA(1) process can be rewritten as an infinite AR process. This is why we rarely work with MA or AR orders higher than 2; most processes can be modeled parsimoniously with combinations of lower-order MA and AR terms. Doing so is not merely indulging an aesthetic favoring simplicity; it reduces the number of coefficients we need to estimate permitting greater precision.

More generally, AR processes have MA representations (and vice versa) as long as the bounds of stationarity are satisfied. Once we have made the appropriate transformation to obtain a stationary series, then, we can model its autocorrelation with MA terms, AR terms, or some combination of the two.

4.6.4 The ACF for an AR(1) Process

To simplify the math and with no loss of generality, we will begin with an AR(1) series where the constant term $\theta_0 = 0$.

$$Y_t = \phi Y_{t-1} + a_t$$

and derive a few of the autocovariances. This will require a couple new tricks.

$$\begin{aligned} \gamma(0) &= E[Y_t^2] \\ &= E[(\phi Y_{t-1} + a_t)^2] \\ &= E[\phi^2 Y_{t-1}^2 + 2\phi Y_{t-1} a_t + a_t^2] \end{aligned}$$

Since $E[Y_{t-1} a_t] = 0$ by construction, the middle term drops out. Also, by the assumption of weak stationarity,

$$Var(Y_t) = E[Y_t^2] = E[Y_{t-1}^2] = \cdots = E[Y_{t-k}^2]$$

for all $k = 0, 1, 2 \ldots t$.

Then our last expression can be rewritten as

$$\begin{aligned} \gamma_0 &= E[\phi^2 Y_{t-1}^2] + E[a_t^2] \\ &= \phi^2 \gamma(0) + \sigma^2 \end{aligned}$$

Rearranging,

$$\gamma(0) = \frac{\sigma^2}{1-\phi^2}$$

$$\begin{aligned}\gamma(1) &= E[Y_t Y_{t-1}] \\ &= E[(\phi Y_{t-1} + a_t)Y_{t-1}] \\ &= E[\phi Y_{t-1}^2 + Y_{t-1}a_t] \\ &= E[\phi Y_{t-1}^2] \\ &= \phi\gamma(0)\end{aligned}$$

$$\begin{aligned}\gamma(2) &= E[Y_t Y_{t-2}] \\ &= E[(\phi Y_{t-1} + a_t)Y_{t-2}]\end{aligned}$$

But remember (lagging the AR(1) equation):

$$Y_{t-1} = \phi Y_{t-2} + a_{t-1}$$

Substituting this into the last equation,

$$\begin{aligned}\gamma(2) &= E[(\phi(\phi Y_{t-2} + a_{t-1}) + a_t)Y_{t-2}] \\ &= E[\phi^2 Y_{t-2}^2] \\ &= \phi^2\gamma(0)\end{aligned}$$

It can be shown that the pattern continues, i.e., for an AR(1) process,

$$\gamma(k) = \phi^k\gamma(0)$$

Recalling that the autocorrelations, $\rho(k)$, can be written

$$\rho(k) = \frac{\gamma(k)}{\gamma(0)}$$

and substituting the expressions we derived for the γ's,

$$\rho(0) = \frac{\gamma(0)}{\gamma(0)} = 1$$

$$\rho(1) = \frac{\gamma(1)}{\gamma(0)} = \frac{\phi\gamma(0)}{\gamma(0)} = \phi$$

$$\rho(2) = \frac{\gamma(2)}{\gamma(0)} = \frac{\phi^2\gamma(0)}{\gamma(0)} = \phi^2$$

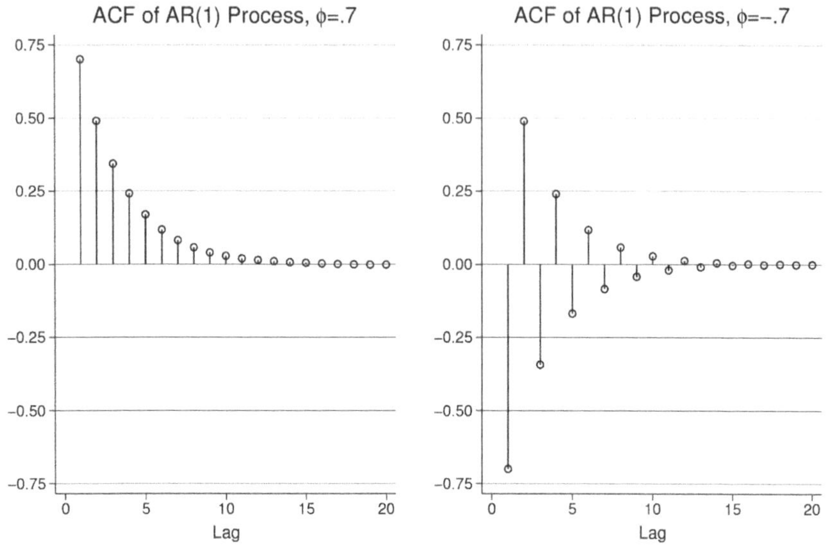

Figure 4.4 ACFs for Two AR(1) Series.

For an AR(1) process,

$$\rho(k) = \phi^k \tag{4.9}$$

The autocorrelation function declines geometrically (but alternates if $\phi < 0$). An illustration is provided in Figure 4.4.

In the example on the left, $\phi_1 = .7$. Given Equation (4.9), what do you expect ρ_2 to be? ρ_3?[4]

In the example on the right, $\phi_1 = -.7$. What do you expect ρ_2 to be? ρ_3?[5]

4.6.4.1 Aside: ACF for a Nonstationary Series

Earlier in the chapter, we described the ACF we would expect to see if we overdifferenced a series. What if we failed to difference a non-stationary series? Notice that our unit root series would be mathematically equivalent to an AR(1) series where $\phi_1 = 1$, with the warning that $\phi_1 = 1$ violates the bounds of stationarity.

$$Y_t = Y_{t-1} + a_t + \theta_0$$

---

4 We know that for an AR(1) series, $\rho_1 = \phi_1$, $\rho_2 = \phi_1^2$, and $\rho_3 = \phi_1^3$. For $\phi_1 = 0.7$, we expect $r_2 = 0.49$ and $r_3 = 0.343$.

5 For $\phi_1 = -0.7$, we expect $r_2 = 0.49$ and $r_3 = -0.343$.

We will use this equation, though, to derive the series' ACF.

$$\begin{aligned}
\rho_1 &= \frac{\mathrm{Cov}(Y_t, Y_{t-1})}{\mathrm{Var}(Y_t)} \\
&= \frac{E[(Y_{t-1} + a_t)Y_{t-1}]}{\mathrm{Var}(Y_t)} \\
&= \frac{E[Y_{t-1}^2 + Y_{t-1}a_t]}{\mathrm{Var}(Y_t)} \\
&= \frac{E[Y_{t-1}^2]}{\mathrm{Var}(Y_t)} \\
&= \frac{\mathrm{Var}(Y_t)}{\mathrm{Var}(Y_t)} \\
&= 1
\end{aligned}$$

In the same fashion, it can be shown that

$$\rho_1 = \rho_2 = \rho_3 \cdots \rho_k = 1$$

As in the left panel of Figure 4.5, the autocorrelation at each k is 1.

Because of the estimation problems associated with non-stationary series, though, the SACF will not look like this. Instead, $r_1 \approx 1$, and the autocorrelations at succeeding lags die out slowly. See the right panel of Figure 4.5 for an illustration.

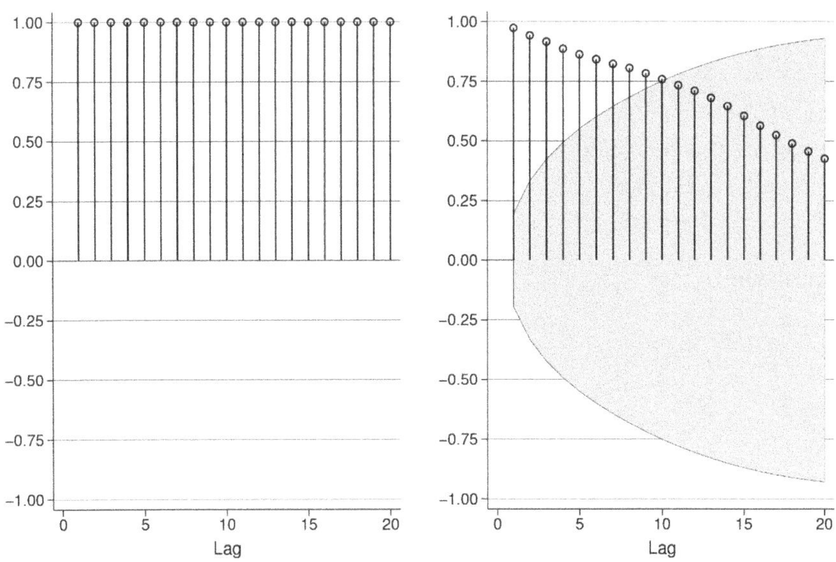

Figure 4.5 ACF and SACF of a Non-Stationary Series.

For an AR(1) series with $\phi_1 < 1$ but very close to 1, it may be difficult to decide whether the series is truly AR(1) or non-stationary; judicious use of unit root tests and the ACF is required. In other cases, two prominent differences between the SACF of an AR(1) series and the SACF of a non-stationary series will be helpful. First, r_1 will not be so close to 1 for most AR(1) series. Second, the SACF of the AR(1) series exhibits geometric decay, but the SACF of the non-stationary series does not.

4.6.4.2 ACF for an AR(2) Process

Once more, to simplify the derivation and without loss of generality, we'll begin with an AR(2) process for which $\theta_0 = 0$.

$$Y_t = \phi_1 Y_{t-1} + \phi_2 Y_{t-2} + a_t$$

To get a general formula for the covariances, $\gamma(k)$, we'll begin by multiplying through by Y_{t-k} and taking expectations:

$$\begin{aligned}\gamma(k) &= E[Y_t Y_{t-k}] \\ &= \phi_1 E[Y_{t-1} Y_{t-k}] + \phi_2 E[Y_{t-2} Y_{t-k}] + E[a_t Y_{t-k}] \\ &= \phi_1 \gamma(k-1) + \phi_2 \gamma(k-2) + E[a_t Y_{t-k}]\end{aligned}$$

But we know that $E[a_t Y_{t-k}] = 0$ for $k \neq 0$. Also, $\rho(0) = 1$ for all series, since it is just the correlation of the series with itself.

For $k = 1, 2, \ldots,$

$$\gamma(k) = \phi_1 \gamma(k-1) + \phi_2 \gamma(k-2)$$

To obtain a general formula for the autocorrelations, $\rho(k)$, we divide this by γ_0:

$$\frac{\gamma(k)}{\gamma_0} = \frac{\phi_1 \gamma(k-1)}{\gamma_0} + \frac{\phi_2 \gamma(k-2)}{\gamma_0}$$
$$\rho(k) = \phi_1 \rho(k-1) + \phi_2 \rho(k-2)$$

We can use this general formula to derive the ρ's:

$$\begin{aligned}\rho(1) &= \phi_1 \rho(0) + \phi_2 \rho(-1) \\ &= \phi_1 + \phi_2 \rho(1)\end{aligned}$$

Rearranging,

$$\rho(1) = \frac{\phi_1}{1-\phi_2}.$$

Then

$$\rho(2) = \frac{\phi_1^2}{1-\phi_2} + \phi_2$$

$$\rho(3) = \frac{\phi_1^3 + \phi_1\phi_2}{1-\phi_2} + \phi_1\phi_2,$$

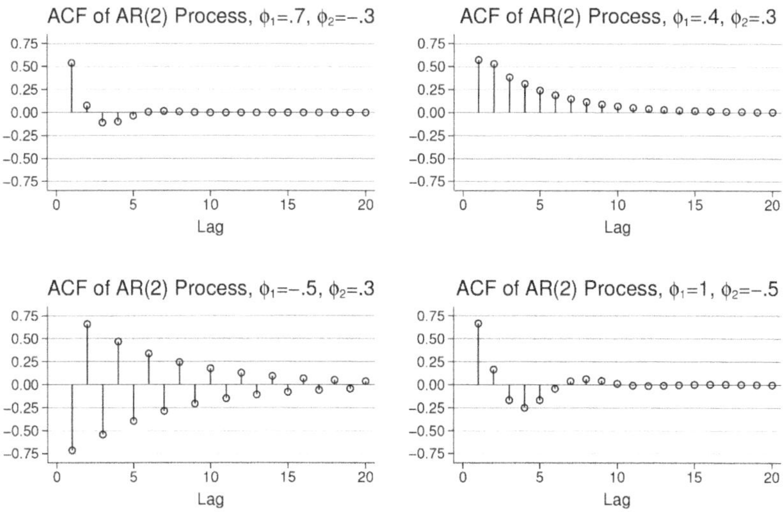

Figure 4.6 ACFs for Four AR(2) Series.

You are not alone if these equations are not painting clear images for you; ACFs of AR(2) processes vary in their presentation. Examples of ACFs for four AR(2) processes are presented in Figure 4.6.

The top right and bottom left examples exhibit decay, but not geometric decay, and alternation for the series with the negative AR(1) term. The series in the top left and bottom right are "wavy," the other common pattern for AR(2) ACFs.

4.6.5 ACF's for AR(p) Processes (General Form)

If you were to attempt the derivation of the ACFs for an AR(3) process, you might begin to notice a pattern. In general,

$$\gamma(k) = \phi_1 \gamma(k-1) + \cdots + \phi_p \gamma(k-p), \quad k > 0 \tag{4.10}$$

$$\rho(k) = \phi_1 \rho(k-1) + \cdots + \phi_p \rho(k-p), \quad k > 0 \tag{4.11}$$

The equations obtained by writing out the first p equations defined by this last equation are referred to as the Yule-Walker Equations. They are useful for deriving expressions for partial autocorrelations, described in the next section.

$$\begin{array}{lclclcl}
\rho(1) & = & \phi_1 & + & \phi_2\rho(1) & + \; \ldots & \phi_p\rho(p-1) \\
\rho(2) & = & \phi_1\rho(1) & + & \phi_2 & + \; \ldots & \phi_p\rho(p-2) \\
 & \vdots & & & & & \\
\rho(p) & = & \phi_1\rho(p-1) & + & \phi_2\rho(p-2) & + \; \ldots & \phi_p
\end{array}$$

As we saw in the ACFs for AR(2) series in Figure 4.6, it can be difficult to distinguish AR(1) processes from AR(2) processes based on their ACFs alone. Discerning between AR(2) and higher-order AR processes based on the ACF is even trickier, as all will exhibit decay. A second diagnostic tool, the Partial Autocorrelation Function (PACF), helps identify an appropriate model when the ACF is ambiguous.

4.7 The Partial Autocorrelation Function

4.7.1 Definitions

The PACF(k) is a measure of the correlation between the series and itself lagged k periods, controlling for all intermediate lags.

Consider the following set of equations:

$$\begin{aligned} Y_t &= \phi_{11} Y_{t-1} + a_{1t} \\ Y_t &= \phi_{12} Y_{t-1} + \phi_{22} Y_{t-2} + a_{2t} \\ &\vdots \\ Y_t &= \phi_{1k} Y_{t-1} + \phi_{2k} Y_{t-2} + \cdots + \phi_{kk} Y_{t-k} + a_{kt} \end{aligned}$$

The partial autocorrelations for the series Y_t are the coefficients $\phi_{11}, \phi_{22}, \ldots, \phi_{kk}$.

$$\text{SE(PACF)} \approx \sqrt{1/n}$$

The general relationship between the partial autocorrelation coefficients, $\phi_{11}, \ldots \phi_{kk}$, and the autocorrelation coefficients, $\rho(1)$, $\rho(2), \rho(k)$ is laborious to derive, but nicely described in Mills (1990: 78-80). For the first three partials, the matrix form can be solved as follows:

$$\phi_{11} = \rho(1) \tag{4.12}$$

$$\phi_{22} = \frac{\rho(2) - [\rho(1)]^2}{1 - [\rho(1)]^2} \tag{4.13}$$

$$\phi_{33} = \frac{-2\rho(1)\rho(2) - [\rho(1)]^2 + [\rho(1)]^3}{1 + 2[\rho(1)]^2 \rho(2) - [\rho(2)]^2 - [\rho(1)]^2} \tag{4.14}$$

We will use these equations to derive PACFs for autoregressive processes.

4.7.2 PACFs for AR Processes

4.7.2.1 PACF for an AR(1) Process

First, from Equation (4.12), we know that for any AR process, the first partial autocorrelation ϕ_{11} is equal to the first autocorrelation $\rho(1)$. Since for an AR(1) process, $\rho(1) = \phi_1$, the coefficient on the Y_{t-1} term,

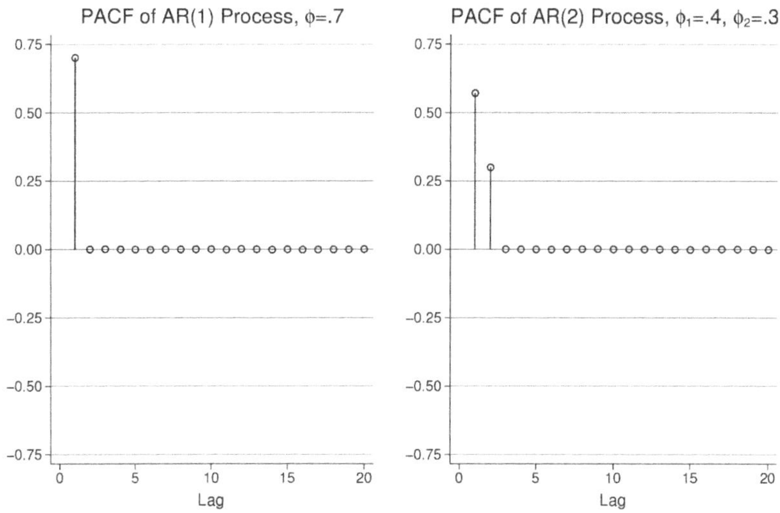

Figure 4.7 PACFs for an AR(1) and an AR(2) Series.

$$\phi_{11} = \phi_1$$

that is, the first partial autocorrelation coefficient will equal the first autocorrelation coefficient.

To obtain the subsequent partial autocorrelations, recall that for an AR(1) series, $\rho(k) = \rho_1^k$. Then the numerators in the subsequent Yule-Walker equations become 0. To summarize,

$$\begin{aligned}
\phi_{11} &= \phi_1 = \rho_1 \\
\phi_{22} &= 0 \\
\phi_{33} &= 0 \\
&\vdots \\
\phi_{kk} &= 0, \quad k > 1
\end{aligned}$$

For an AR(1) process, the PACF has a spike at lag 1, and then goes immediately to 0. It is no coincidence that the PACF for an AR(1) process exhibits a pattern similar to the ACF for an MA(1) process; it is another consequence of the duality between AR and MA processes. An example is provided in the left panel of Figure 4.7.

4.7.2.2 PACF for an AR(2) Process

In similar fashion, we can derive the partial autocorrelations ϕ_{kk} by beginning with Equations (4.12–4.14), and then substituting the expressions for the autocorrelations, $\rho(k)$, that we derived earlier in the chapter.

$$
\begin{aligned}
\phi_{11} &= \frac{\phi_1}{1-\phi_2} \\
\phi_{22} &= \frac{\phi_2(\phi_2-1)^2-\phi_1\phi_2}{(1-\phi_2)^2-\phi_1^2} \\
\phi_{33} &= 0 \\
&\vdots \\
\phi(k) &= 0,\ k>2
\end{aligned}
$$

That is, for an AR(2) process, the PACF has spikes at lags 1 and 2, and then goes immediately to 0. See the right panel of Figure 4.7 for an example.

The pattern continues. For AR(p) series, the PACF will have spikes through lag p, and then drop immediately to 0 for lags k > p.

4.7.3 PACFs for MA Processes

Derivations of the PACFs for MA processes are quite messy, and we won't go through them here. Their shape should be unsurprising at this point, given what you know about the duality between AR and MA processes and about the shape of PACFs for AR processes: PACFs for MA(q) series have shapes like the ACFs for AR(q) series. That is, the PACF for an MA(1) series will show geometric decay like the ACF of an AR(1) series, alternating if the MA coefficient is negative. The PACF for an MA(2) series will exhibit either decay or "waves." The Figure 4.8 provides examples of ACFs and corresponding PACFs for two MA and three AR series.

4.8 Seasonality

Many time series exhibit seasonal patterns – recurring patterns at regular time intervals such as day of the week, season of the year, or in countries with regularly scheduled elections, year in the election cycle. In Chapter 3, we examined one such series – monthly averages of the concentration of atmospheric carbon dioxide (CO_2) from 2006 through 2012. The raw series, shown in Figure 3.10, exhibits a clear cyclical pattern with increases through the North's Fall and Winter as leaves are lost and photosynthesis ebbs, and declines through the North's Spring and Summer when increasing foliage absorbs increasing amounts of CO_2.

Cases such as this exhibit seasonal autocorrelation. In monthly data, we expect correlation between Y_t and Y_{t-12}; in quarterly data we expect correlation between Y_t and Y_{t-4}, in daily data we might expect correlation between Y_t and Y_{t-7}.

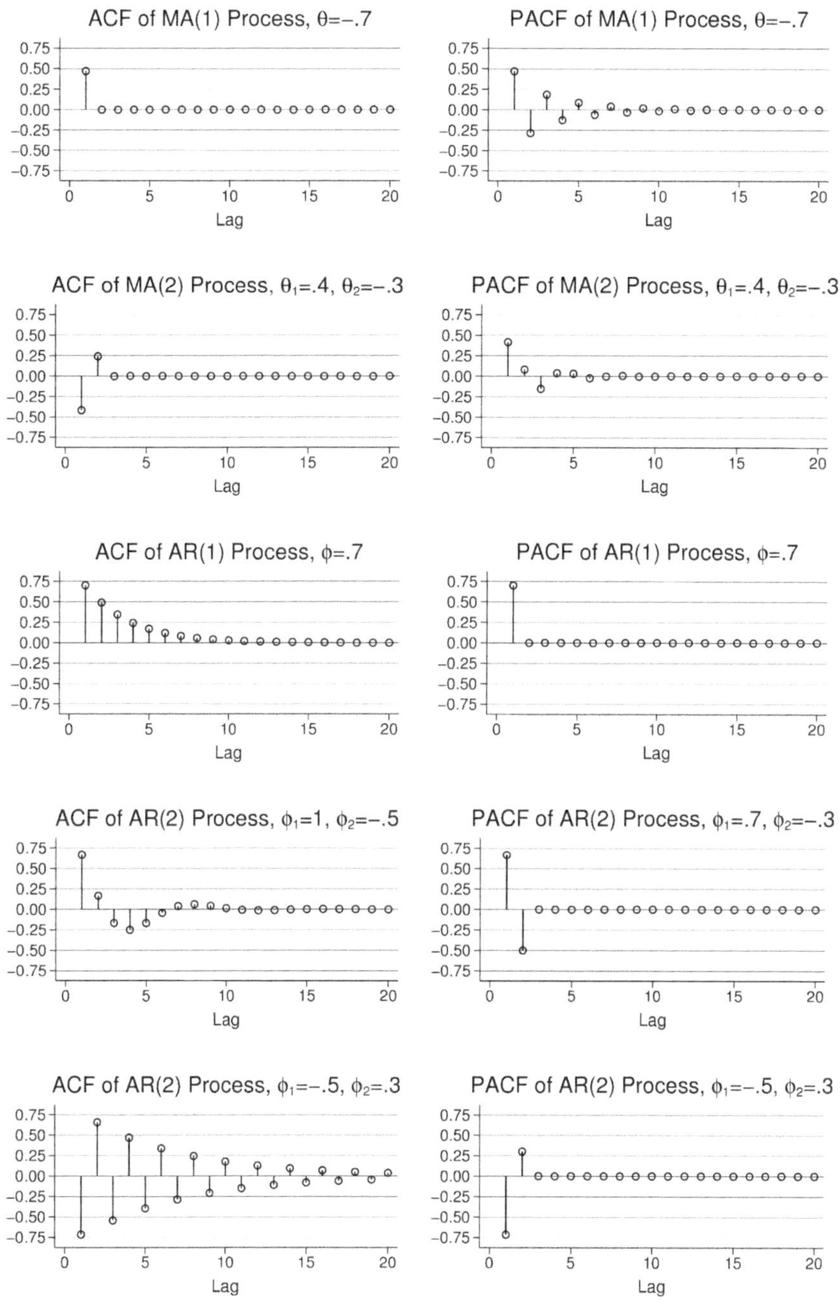

Figure 4.8 ACFs and PACFs for MA and AR Processes.

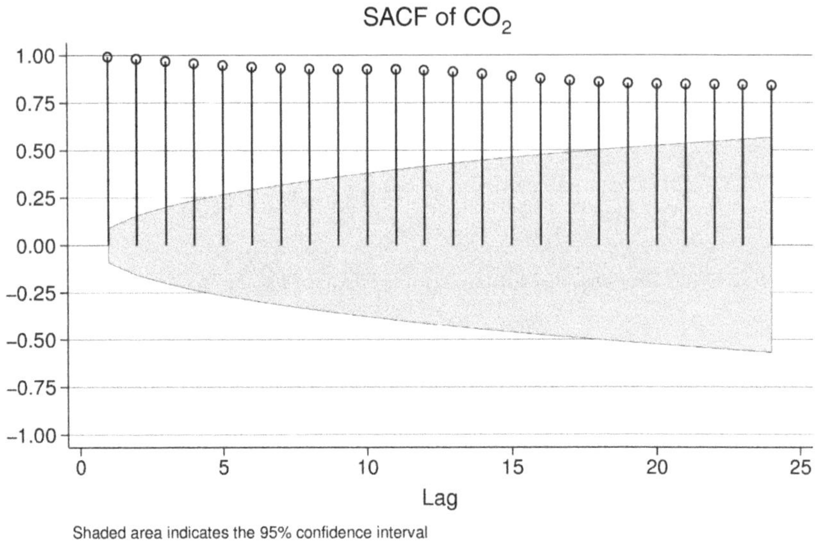

Figure 4.9 ACF for the Raw Atmospheric CO_2 Series.

There is a further complication in the CO_2 series: it is clearly increasing over time. Not surprisingly, the ACF for the raw series, shown in Figure 4.9, indicates clear non-stationarity.

The ACF for the raw atmospheric CO_2 series is shown in Figure 4.9.

Reflexively differencing or de-trending the series, though, would be a mistake. In the case of non-stationary series with seasonal effects, the seasonal component and non-stationarity are best addressed simultaneously.

4.8.1 Seasonal Non-stationarity

In principle, seasonal non-stationarity occurs when there is a perfect correlation between a series Y_t and its value at its seasonal lag:

$$Y_t = \theta_0 + Y_{t-S} + a_t$$

where S indicates the number of time points in a season. In the case of quarterly data,

$$Y_t = \theta_0 + Y_{t-4} + a_t$$

To achieve stationarity, difference the season seasonally. In the case of quarterly data,

$$\begin{aligned} Y_t - Y_{t-4} &= \theta_0 + a_t \\ \Delta_{S4} Y_t &= \theta_0 + a_t \end{aligned}$$

Table 4.2 Illustration of Seasonally Differencing a Quarterly Series with a Trend.

X_t	S_t	Z_t	$Z_t - Z_{t-4}$
1	+1	2	–
2	+2	4	–
3	−1	2	–
4	−2	2	–
5	+1	6	4
6	+2	8	4
7	−1	6	4
8	−2	6	4
9	+1	10	4
10	+2	12	4
11	−1	10	4
12	−2	10	4

Notice that the seasonal difference is not the same as the fourth difference, $\Delta^4 Y_t$

$$\begin{aligned}
\Delta^4 Y_t &= (1-L)^4 Y_t \\
&= (1-L)^2(1-L)^2 Y_t \\
&= (1-2L+L^2)(1-2L+L^2)Y_t \\
&= (1-4L+6L^2-4L^3+L^4)Y_t \\
&= Y_t - 4Y_{t-1} + 6Y_{t-2} - 4Y_{t-3} + Y_{t-4}
\end{aligned}$$

In the case of seasonal data, we are not concerned that the non-stationarity might be driven by a deterministic trend because seasonally differencing the data will still be the correct transformation. To see this, consider the contrived quarterly data shown in Table 4.2.[6]

Xt is a linear trend; it increases by one point each time period. *St* represents a seasonal effect: 1 is added in each first quarter; 2 is added in each second quarter; 1 is subtracted in each third quarter, and 2 is subtracted in each fourth quarter. This yields our series Z_t, constructed from a linear trend with a (quarterly) seasonal component. The last column in Table 4.2 lists the values of the quarterly-differenced $Z-t$ series, $\Delta_{S4}Z_t$. Seasonally differencing the series constructed from a deterministic trend and a seasonal component has left us with a stationary series.

6 This illustration is a revised version of a brilliant example provided by Charemza and Deadman, 1997:37-38.

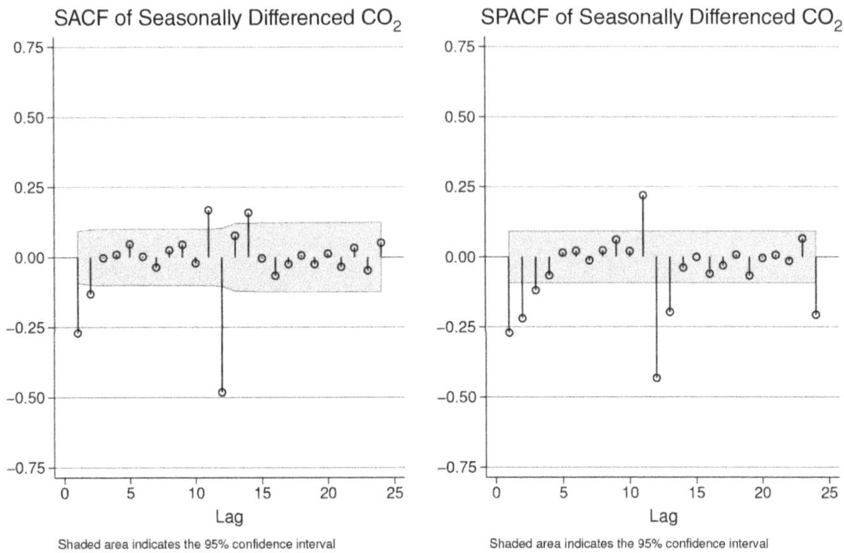

Figure 4.10 SACF and SPACF Seasonally Differenced CO_2.

The ACF and PACF for the seasonally differenced CO_2 series is presented in Figure 4.10.

Notice that the characteristic mark of a nonstationary series – autocorrelations near one that damp out only slowly – has disappeared after the transformation. Notice also that there are large spikes in both the SACF and SPACF at lag 12 (and smaller spikes nearby) in addition to spikes in the early lags. This evidence points to a remaining seasonal component, as well as a regular MA and/or AR component. Modeling such a complex autocorrelation structure will require estimating candidate models, examining the residuals, and performing other diagnostics. We will turn to this topic in Chapter 5, but first let's discuss the possibilities for the remaining seasonal component.

4.8.2 Seasonal Moving Average

In the same way that it is possible to have seasonal non-stationarity, it is possible to have a seasonal MA component. In the case of quarterly data, a seasonal MA(1) process takes the form

$$Y_t = \theta_0 + a_t - \theta a_{t-4}$$

or

$$Y_t = \theta_0 + (1 - \theta L^4) a_t.$$

Its ACF and PACF look exactly like those of a regular MA(1) process, but the resulting spikes will occur only at seasonal lags – e.g., at 4, 8, 12... for a quarterly series, 12,24,36... for a monthly series, etc. The logic can be extended to higher-order seasonal MA processes. Notice that if our season covers many periods – e.g., 12 periods for monthly data – and our time series is not very long, we may not be able to produce an SACF and SPACF with enough lags to see a clear pattern. This is an additional reason why estimating candidate models and using diagnostic critieria to decide which we prefer becomes necessary.

4.8.3 Seasonal Autoregression

Similarly, a seasonal AR(1) process for quarterly data takes the general form

$$Y_t = \theta_0 + \phi Y_{t-4} + a_t$$

or

$$(1 - \phi L^4)Y_t = a_t.$$

Its ACF and PACF look exactly like those of a regular AR(1) process, but the resulting spikes will occur at the seasonal lags – e.g., at 4,8,12... for the quarterly series. The logic can be extended to higher-order AR processes, and the same caveat about the length of our season compared to the length of our data applies. If a single season covers many time points and we do not have a long time series, we may not be able to produce an SACF and SPACF with enough lags to see a clear pattern.

4.8.4 Combining Regular and Seasonal Components

In Chapter 5, you will learn that worrying about spikes that don't have an apparent explanation is typically unnecessary; they often disappear after we correctly model the piece of the autocorrelation structure that we *can* guess from the SACF and PACF. Examining the equation for model that has both a regular component and a seasonal component may help convey some intuition.

The general form for monthly data with both a regular AR(1) and seasonal AR(1) can be written as follows:

$$(1 - \phi_1 L)(1 - \phi_{12} L^{12})Y_t = \theta_0 + a_t$$

$$(1 - \phi_1 L - \phi_{12} L^{12} + \phi_1 \phi_{12} L^{13})Y_t = \theta_0 + a_t$$

$$Y_t = \theta_0 + \phi_1 Y_{t-1} + \phi_{12} Y_{t-12} + \phi_1 \phi_{12} Y_{t-13} + a_t$$

Notice the appearance of Y_{t-13}. Its presence will result in a spike in the ACF at the 13th lag.

4.9 ARMA (mixed) Processes

Finally, just as we may have a mix of regular and seasonal components, we may have a mix of regular AR and regular MA components. Both the ACFs and PACFs of these series show decay, and we will need to estimate candidate series in order to identify them. We will turn to this topic in Chapter 5.

4.10 Summary

- Autocorrelation is a property of most time series, and should be carefully modeled.
- Autocorrelation in our dependent variable that is not addressed by independent variables will appear in the error term, where it may introduce omitted variables bias into our estimated coefficients.
- Wold's Theorem provides a justification for modeling any weakly stationary purely nondeterministic stochastic process as a moving average (MA) process.
- MA(q) processes take the form $Y_t = \theta_0 + a_t + \theta_1 a_{t-1} + \theta_2 a_{t-2} + \cdots + \theta_q a_{t-q}$.
- Autocorrelation functions (ACFs) summarize the correlations $\rho(k)$ between a theoretical series and lags of itself. Sample autocorrelation functions (SACFs) summarize the correlations $r(k)$ between a particular realization and lags of itself.
- The ACF of an MA(q) series has spikes up to lag q, then drops sharply to 0.
- Autoregressive (AR) processes take the form $Y_t = \theta_0 + \phi_1 Y_{t-1} + \phi_2 Y_{t-2} + \cdots + \phi_p Y_{t-p}$.
- Because of the duality between MA and AR processes, series can typically be modeled with low-order MA and/or low-order AR components.
- ACFs of AR(1) series exhibit geometric decay. If $\phi_1 < 0$, the spikes will alternate between positive and negative.
- ACFS of AR(2) series may decay or exhibit waves, and can be hard to interpret.
- Partial autocorrelation functions (PACFs) are a second diagnostic tool, particularly useful for discerning between AR(2) and MA(2) processes. As a consequence of the duality between AR and MA processes, PACFs of AR processess look like ACFs of MA processes, and vice versa.

- Seasonal differencing can be used to address deterministic trends as well as stochastic trends, and should be considered when seasonality is observed before differencing a nonstationary series.
- Seasonal AR and MA exhibit the same patterns in the SACF and SPACF, but at the seasonal lags (e.g., 4, 8, 12 etc. for quarterly data).
- ARMA models, models with both AR and MA terms, cannot be identified based on the SACF and SPACF alone; estimation is required. Methods for estimating AR and MA and combined processes will be covered in Chapter 5.

References

Charemza W. and Deadman D. (1997). *New Directions in Econometric Practice: General to Specific Modelling, Cointegration and Vector Autoregression,* 2nd ed. Cheltenham, UK: Edward Elgar.

Mills T.C. (1990). *Time Series Techniques for Economists.* Cambridge: Cambridge University Press.

5

Autocorrelation: Univariate ARIMA Estimation and Forecasting

Chapter 4 provided a reminder that autocorrelation is a property of most time series. In the same way that two series which are both non-stationary will be correlated with or without any reasonable causal mechanism, two time series with a common autocorrelation structure are often correlated even if there is no plausible relationship between them. This should be no surprise – remember that autocorrelation is often driven by omitted variables. Appropriately addressing autocorrelation, then, is not merely as important as addressing confounding variables in order to prevent omitted variable bias; it is frequently the same thing.

Different time series techniques address the problem in different ways. Beginning with Chapter 9, we'll see that recent approaches favor modeling autocorrelation by including enough lags of the independent and/or dependent variables to eliminate any autocorrelation in the error terms. As troubling as this strategy often seems at first to well-trained statisticians accustomed to working with cross-sectional data, it serves two purposes. First, it prevents the omitted variables bias that might arise from autocorrelation driven by omitted variables. Second, it facilitates the assessment of dynamic effects – the response of our dependent variable, over time, to a change in the independent variable(s) of interest. A preview of this sort of analysis will appear in the discussion of transfer functions in Chapter 6, and will recur in the context of dynamic specification in chapter 9 and the impulse response functions derived from vector autoregression models in chapter 12.

This chapter introduces the autoregressive integrated moving average (ARIMA) model-building process popularized by Box and Jenkins (1970). ARIMA models use either or both autoregressive (AR) and moving average (MA) terms to model autocorrelation. ARIMA estimation is a useful tool for several reasons.

Applied Time Series Analysis for the Social Sciences: Specification, Estimation, and Inference, First Edition. Regina M. Baker.

Companion website: www.wiley.com/go/baker

- As we will see in the identification section that follows, sample autocorrelations (SACFs) are not always as clear as the ACFs presented in Chapter 4. The ARIMA estimation process helps clarify the nature of the underlying autocorrelation whether or not an ARIMA model is the long-run goal. Indeed, the LSE approach introduced in Chapter 11 makes use of both standard ARIMA diagnostics and forecasting in its model-building. ARIMA estimation can be especially helpful for confirming the nature of seasonal effects.
- Although vector autoregression forecasts outperform univariate ARIMA forecasts – especially in the longer term – univariate ARIMA models require less information and were found to provide better forecasts in many settings than much more complex, theory-driven simultaneous equations models.
- By themselves, univariate ARIMA models can be useful for providing an atheoretical baseline for examining simple intervention effects – that is, questions like "Did a policy change have an effect?" Models such as this are sometimes referred to as "interrupted time series" models and will be explored in Chapter 6.
- ARIMA models that include a continuous variable are sometimes referred to as "multivariate ARIMA" models in the ARIMA literature, and will be covered in Chapter 7. Through a process called "prewhitening," the dependent and independent variables are purged of their common autocorrelation structure to mitigate the concern that any shared component will lead to spurious results. The generalized least squares (GLS) models that are introduced in introductory regression courses as a means for correcting first-order autocorrelation are a special case of multivariate ARIMA based on the assumption that the error terms follow an AR(1) process, but by now, you should suspect that not all error structures are AR(1). ARIMA estimation permits estimation of models with more complex autocorrelation.

The chapter proceeds as follows. A discussion of ARIMA notation will be followed by an introduction to the ARIMA model-building process. The SACFs and SPACFs presented in Chapter 4 provide the basis for the first step, identification; the estimation, diagnosis and metadiagnosis steps are discussed in turn. The process will be illustrated using the pre-intervention portion of a series that will be fully modeled in Chapter 6. The final section will cover forecasting in ARIMA models.

5.1 ARIMA $(p,d,q)(P,D,Q)_s$ Notation

In Chapter 4, we learned from Wold's Theorem that every weakly stationary non-deterministic stochastic process can be written as a moving average (MA) process, and further, that MA processes can be rewritten in autoregressive

(AR) form. As a practical matter, this is useful because it permits us to model autocorrelation using fewer terms – e.g., rewriting a high-order MA process as a lower-order AR process so that fewer parameters need to be estimated. We limited our attention, though, to models that had only AR or only MA components and no seasonal components. Sometimes, however, autocorrelation is best modeled by a combination of AR and MA terms. ARIMA models are designed to do exactly that – model a series using a relatively parsimonious combination of AR and MA components.

$$\Delta^d Y_t = \theta_0 + \phi_1 Y_{t-1} + \phi_2 Y_{t-2} + \ldots + \phi_p Y_{t-p} + a_t - \theta_1 a_{t-1} - \theta_2 a_{t-2} - \ldots - \theta_q a_{t-q}. \quad (5.1)$$

Models without seasonal components are neatly summarized by three numbers, p, d, and q, taken from the general form of a non-seasonal ARIMA model shown in Equation (5.1), where

- a_t is our white noise error term,
- θ_0 is a constant term,
- ϕ_k is the coefficient on the AR term at lag k,
- $\theta_k, k \neq 0$, is the coefficient on the MA term at lag k,
- p represents the highest-order AR component,
- d represents the order of differencing used to make the original series stationary, and
- q represents the highest-order MA component.

So, for example, ARIMA (1,0,0) is the full notation for the set of models presented in Section 4.6.1; their highest AR order was 1, the series were not differenced, and there was no MA component. They can be written as follows:

$$Y_t = \theta_0 + \phi_1 Y_{t-1} + a_t$$

Alternatively, (0,1,1) is notation for the following model:

$$\Delta Y_t = \theta_0 + a_t - \theta_1 a_{t-1}$$

When using this shorthand with statistical software, it is important to check the documentation carefully for the form of the equation used and to notice the signs preceding the coefficients. Equation (5.1) is standard notation in most textbooks, and with some statistical software. When models are parameterized this way, though, estimates of MA terms shown in the results will have the opposite sign from the way they appear in the equation; e.g., a coefficient estimate of 0.42 for an MA(1) term in an ARIMA (0,0,1) model would indicate the following equation:

$$Y_t = \theta_0 + a_t - 0.42 a_{t-1}.$$

If, however, your software documentation reports that it estimates the following model:

$$\Delta^d Y_t = \theta_0 + \phi_1 Y_{t-1} + \phi_2 Y_{t-2} + \ldots + \phi_p Y_{t-p} + a_t + \theta_1 a_{t-1} + \theta_2 a_{t-2} + \ldots + \theta_q a_{t-q}. \quad (5.2)$$

the same coefficient estimate reported for an ARIMA (0,0,1) model would be written as follows:

$$Y_t = \theta_0 + a_t + .42a_{t-1}.$$

Finally, in section 3.7, we saw that series sometimes have seasonal components. For example, atmospheric CO_2 concentrations tend to be higher in winter months because of the sparser foliage absorbing it. In quarterly data, then, we expect to see correlations between observations four periods apart; in monthly data, we expect to see correlations between observations twelve periods apart. These effects are modeled separately, and indicated with additional notation as ARIMA(p,d,q) $(P,D,Q)_S$ models, where

- P represents the highest-order seasonal AR component,
- D represents the order of seasonal differencing needed to make the original series stationary,
- Q represents the highest-order seasonal MA component, and
- S represents the number of periods in a full cycle; for example, $S = 4$ for quarterly data; $S = 12$ for monthly data.

The notation $(0,1,0)(1,0,0)_4$, for example, represents a model in which a series reported quarterly is first-differenced, and includes only a seasonal AR(1) term. It could be written as follows:

$$\Delta Y_t = \theta_0 + \phi Y_{t-4} + a_t.$$

5.2 The ARIMA Model-Building Process

Building ARIMA models entails an iterative process of making sure the series is stationary (transforming if necessary), preliminary identification of any AR and MA components using the ACF's and PACF's presented in Chapter 4, estimation of the preliminary model, and diagnosis of the estimated model. If more than one model satisfies the criteria at the diagnostic stage, metadiagnostic tools can be used to decide between them.

The general process is sketched in Figure 5.1, and each step is described in detail in the sections that follow.

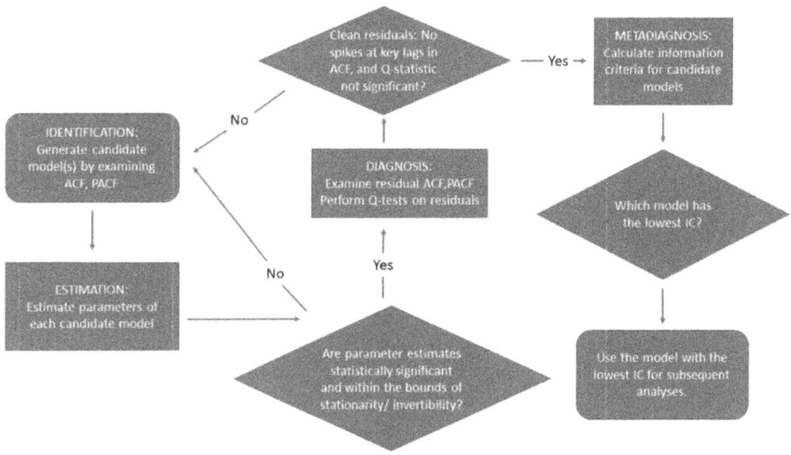

Figure 5.1 The ARIMA model-building process.

5.2.1 Identification

Chapter 4 introduced the tools used for preliminary identification of the autocorrelation structure of time series – sample autocorrelation functions (SACFs) and partial autocorrelation functions (SPACFs).

A quick review of Figure 5.2 might convince you that SACFs are often inadequate, by themselves, for finding the best model of the autocorrelation. All four SACFs were simulated as AR(1) DGPs with $\phi = 0.5$, $n = 100$, but with four different starting points and randomly-generated error terms. In the two on the left panel, only the first autocorrelation is statistically significant; we might easily guess this was an MA(1) process rather than an AR(1). The first few lags in the two SACFs in the right panel look reasonably like geometric decay, as we would expect for an AR(1) process, but only the first two spikes in each are statistically significant. We might also consider the possibility that they are MA(2) processes.

The first step in building an ARIMA model, then, is to develop a list of candidate models. Start with the most parsimonious. For example, for the models in the left panel, (1,0,0) and (0,0,1) models are better starting points than a (1,0,1) model for two reasons. First, an inappropriately included term can "induce" a false positive for another term, in the same way that over-differencing can induce an MA(1) term. Second, because regular (versus seasonal) ARIMA components are additive, any component that is missed in the first estimation will be easier to identify by looking at the residuals from the estimation. This will become clearer as we work through examples.

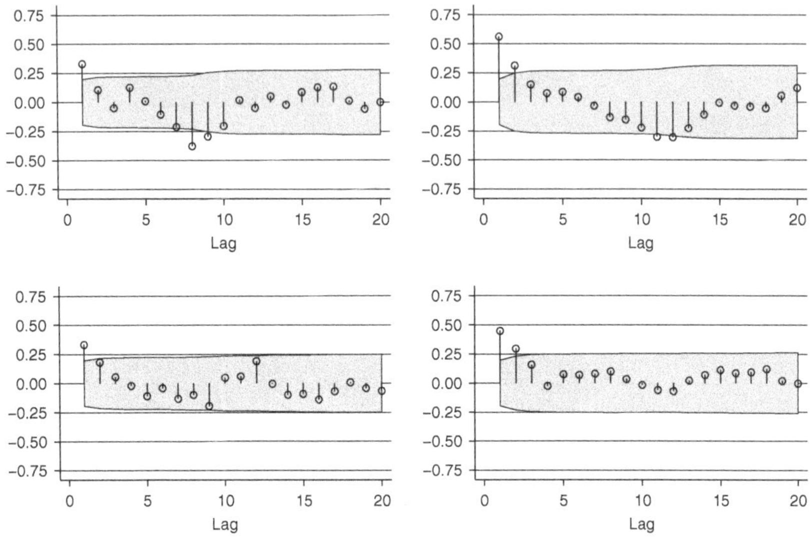

Figure 5.2 SACFs from Simulated AR(1) Series $Y_t = 0.5Y_{t-1} + a_t$, $a_t \sim N(0,1)$, and $n = 100$ for all four series. The shaded areas represent 95% confidence bounds.

5.2.2 Estimation Methods

Simple regression models can be estimated using ordinary least squares (OLS). Maximizing the sum of the squared error terms leaves us with an equation that can be solved directly for the estimate of the coefficient of interest. In the case of ARIMA models, however, this is not possible because the maximization results in an equation which is nonlinear in the parameters of interest – the ρs in AR models and/or the θs in MA models. The mathematical details are presented in the appendix to this chapter. The upshot, though, is that alternative methods – numerical estimation techniques – must be used. Briefly, numerical estimation techniques involve trying different hypothesized values for the parameter estimates, and then evaluating the fit of the model using some criterion. Further details are presented in the appendix. The three options most commonly available for ARIMA estimation in standard statistical software are maximum likelihood estimation (MLE), unconditional least squares (ULS) and conditional least squares (CLS):

- MLE involves maximizing the likelihood of the data given the hypothesized model, and is usually the preferred method.

- ULS uses an iterative algorithm, but the criterion for fit is the minimized sum of the squared errors instead of the maximized likelihood function.
- CLS uses an approximation to the likelihood function that is less computationally intensive, but not as accurate. It was more commonly used decades ago when computing power was much more limited than it is now.

5.2.3 Diagnosis

After estimating a model, estimated coefficients are checked to ensure that they are statistically significant and within the bounds of stationarity, for AR components, and/or invertibility, for MA components, as described in Chapter 4. The bounds for seasonal components are the same as those for regular terms. For example, if we are working with quarterly data and estimate the following SAR(1) model:

$$Y_t = \phi Y_{t-4} + a_t$$

the bounds of stationarity are met if $-1 < \phi < 1$.

If either of these criteria is not met, we return to the identification stage. Otherwise, we proceed to analyze the estimated error terms.

To decide whether there is autocorrelation remaining in the estimated model residuals, we examine the SACF of the residuals and compute the associated Q-tests. There should be no significant spikes at key lags – i.e., the first couple of lags and those associated with seasonal components. Recall from Chapter 4 that the $Q(k)$ statistic allows us to test the null hypothesis that our DGP autocorrelations through lag k are zero. The choice of k is somewhat arbitrary, but in the case of seasonal data, should cover at least one time point past a full season. Covering more than one season is preferable, if the series is long enough. We should not be able to reject the null hypothesis associated with the Q-test.

If all of these criteria are satisfied, we have an adequate model. If not, we examine the SACF and SPACF of the residuals for clues to the remaining components that need to be included.

5.2.4 Meta-diagnosis: the Use of Information Criteria

It is not uncommon to find that more than one ARIMA specification is adequate – i.e., that at least two different models meet the criteria specified in the "diagnosis" step. As in other contexts, information criteria can be used to decide between models. In general, information criteria are calculated to reward models that fit the data better, but penalize models for including additional terms. The two most widely available in popular software are the Akaike

Information Criterion (AIC) and Bayesian information criterion (BIC). They can be calculated as

$$\text{AIC} = -2\ln(\text{likelihood}) + 2k$$

$$\text{BIC} = -2\ln(\text{likelihood}) + k\ln(T)$$

where

- k is the number of estimated parameters, and
- T is the number of time points.

Lower AIC and BIC indicate better models. Since $ln(T) > 2$ for all practical applications, the BIC has a higher penalty for including additional terms.

There are various formulae used in the literature, though, and by different statistical packages. It is not uncommon, for example, to see these formulae divided by the number of cases used. As a consequence, one must be careful about comparing information criteria computed by different statistical packages.

5.3 Example: Directory Assistance (411)

One of the most useful applications of ARIMA estimation is intervention analysis – assessing the impact of an intervention such as a policy change on an outcome of interest. In their classic *Applied Time Series Analysis for the Social Sciences*, McCleary and Hay presented an ARIMA intervention model of the number of calls to directory assistance ("411" in the US) in Cincinnati, Ohio.[1]

A time series plot of the raw series is presented in Figure 5.3.

The number increased over the long run until the telephone company began charging for these calls in March of 1974; subsequently there was a precipitous drop. In this case, the change is dramatic and there should be little doubt that the change is not simply part of the random variation of the series. In many other cases, though, this is less clear, and ARIMA intervention models can be used to discern between the two possibilities.

The first step in assessing the impact is to model the baseline variation in the series; we will do that in this chapter, and model the intervention in the next chapter. In cases where the effect of the intervention is less pronounced, it is often possible to use the entire series to model the baseline, but in this case, it is necessary to begin by modeling the pre-intervention series separately.

---

1 McCleary & Hay, 1980; McSweeny, 1978. McCleary and Hay were limited by the computational software available at the time. Their results can be replicated by using conditional least squares estimation. Maximum likelihood estimation will be used in this text, and leads to slightly different results.

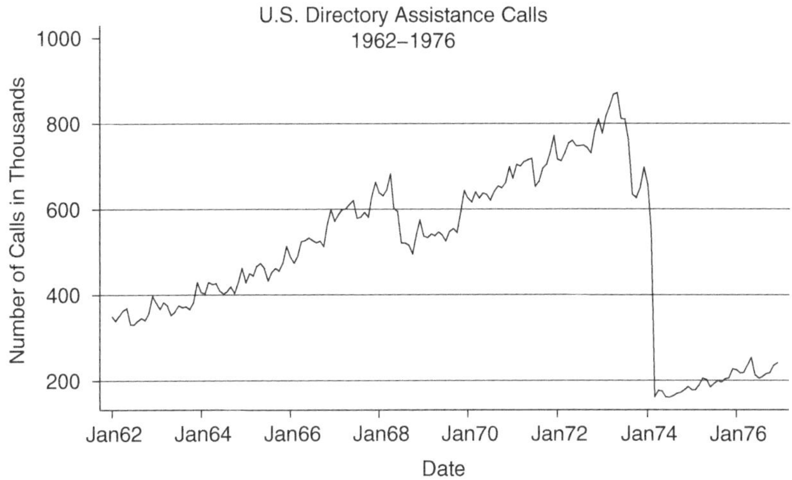

Figure 5.3 Monthly number of calls to directory assistance (411), January 1962–December 1976.

The most obvious feature of the pre-intervention series is its non-stationarity. To decide whether the trend is stochastic or deterministic, we perform an augmented Dickey–Fuller test, and obtain the results presented in Equation (5.3):

$$\Delta Y_t = \underset{(2.3)}{42.9} + \underset{(-2.11)}{-0.110} Y_{t-1} + \underset{(1.60)}{0.278} t + \underset{(0.112)}{-0.094} \Delta Y_{t-1} + a_t, \tag{5.3}$$

where $n = 144$, $a_t \sim N(0,\sigma^2)$, and estimated t-statistics are presented in parentheses. The critical values for the Dickey–Fuller test are –3.444 and –3.144 for the 5% and 10% levels of significance respectively. Since the t-statistic associated with the unit root is –2.11, well above the critical value for the 10% level of significance, we cannot reject the null hypothesis of a unit root at the 0.10 level and will difference the series.

The SACF and SPACF for the first-differenced (0,1,0) series appear in Figure 5.4. Notice that 36 lags were included in both, covering three full seasons.

The SACF and PACF may appear daunting at first. While there are clear seasonal spikes at 12, 24, and 36 in both the SACF and SPACF, there are also puzzling spikes at lags 2, 10, and 22. Inexplicable spikes need not be a concern at this point; you will see that they often disappear when we model the pieces that do make sense – in this case, a seasonal component of some sort. That said, the seasonal spikes do not provide a clear prescription. The three spikes in the SACF are declining, but not necessarily geometrically; a seasonal AR(1) term or seasonal differencing are equally plausible alternatives. Let us start by adding a seasonal

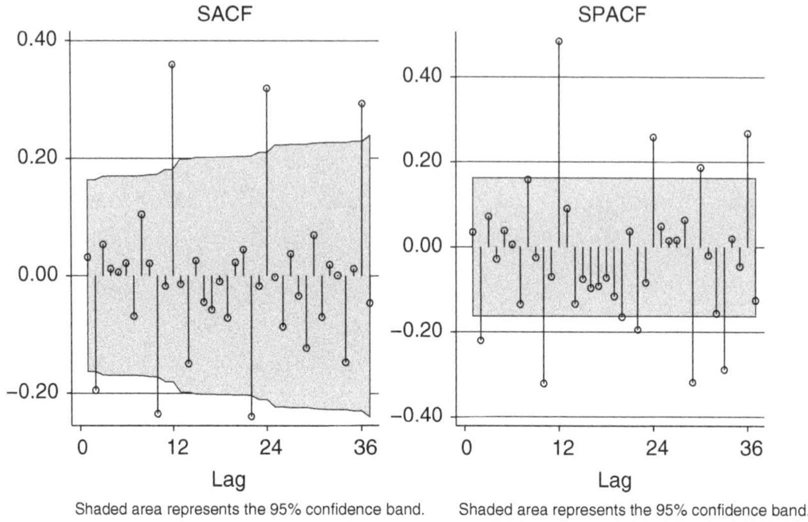

Figure 5.4 The SACF and SPACF for the first-differenced 411 series $(0,1,0)(0,0,0)_{12}$.

Table 5.1 ARIMA estimates for the 411 pre-intervention series.

Model	Parameter	Estimate	Standard Error	*P*-value	AIC	SBC
$(0,1,0)(1,0,0)_{12}$	θ_0	–0.26	5.4	0.96	1361	1370
	φ_{12}	0.51	0.08	< 0.001		
$(0,1,0)(0,1,1)_{12}$	θ_0	–0.78	0.69	0.25		
	θ_{12}	+1.0	456	1.0		
$(0,1,0)(1,1,0)_{12}$	θ_0	−1.46	2.1	0.49	1255	1264
	φ_{12}	–0.44	0.06	< 0.001		

P-values are reported for two-sided hypothesis tests of $\phi_k = 0$ or $\theta_k = 0$.

AR(1) component, estimating the model, and then examining its estimated residuals for more clues.

The estimates for the $(0,1,0)(1,0,0)_{12}$ model appear in the first section of Table 5.1.

The estimated equation can be written as follows:

$$\Delta Calls_t = -0.26 + 0.51 \Delta Calls_{t-12} + a_t,$$

where a_t is our white noise error term.

Recall our criteria for the "diagnosis" stage. First, the estimated coefficients on the AR and/or MA components should be statistically significant and must be within the bounds of stationarity and/or invertibility. $\varphi_{12} = 0.51$ is statistically significant ($p < 0.001$), and within the bounds of stationarity for an AR(1) model: $-1 \le \varphi_{12} \le 1$.

Second, the estimated residuals should appear to be white noise; that is, the SACF of the estimated residuals should not have spikes at key lags, and we should be able to reject the null hypothesis of joint statistical significance of autocorrelations using the Q-test.

The ACF of the model's residuals are shown in Figure 5.5.

There are no significant spikes anywhere in the SACF. Q-tests through lag 24 are presented in Table 5.2. The Q-test at lag 1 tests the null hypothesis that $ACF(1) = 0$. Our p-value of 0.78 indicates that we cannot reject the null hypothesis at any standard level of statistical significance.

The Q-test at lag 2 tests the null hypothesis that $ACF(1) = ACF(2) = 0$ – i.e., that the first two autocorrelations in the DGP are equal to zero. The associated p-value of 0.71 does not allow us to reject the null hypothesis. Indeed, none of Q-tests provide evidence that there is autocorrelation remaining in the residuals. The $(0,1,0)(1,0,0)_{12}$ model, then, passes our diagnostic criteria; it is an adequate model.

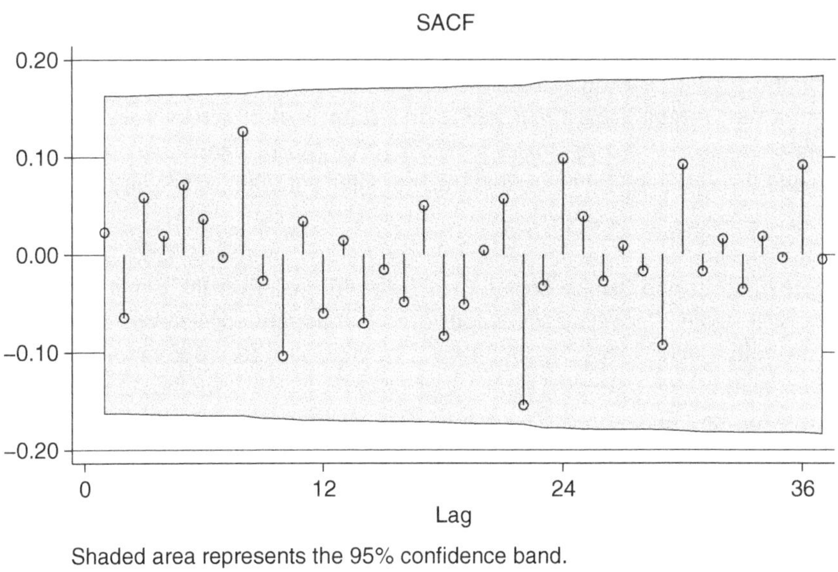

Figure 5.5 The SACF for the residuals from the $(0,1,0)(1,0,0)_{12}$ model.

Table 5.2 Q-tests for the residuals from the ARIMA $(0,1,0)(1,0,0)_{12}$ model

Lag	*Q*	*P*-value	Lag	*Q*	*P*-value
1	0.08	0.78	13	7.35	0.88
2	0.70	0.71	14	8.15	0.88
3	1.21	0.75	15	8.19	0.92
4	1.27	0.87	16	8.57	0.93
5	2.06	0.84	17	9.00	0.94
6	2.27	0.89	18	10.16	0.93
7	2.27	0.94	19	10.60	0.94
8	4.75	0.78	20	10.61	0.96
9	4.86	0.85	21	11.17	0.96
10	6.55	0.77	22	15.30	0.85
11	6.74	0.82	23	15.47	0.88
12	7.32	0.84	24	17.19	0.84

It may not be the only adequate model, though. Remember that we were unsure whether we should include a seasonal AR(1) term or seasonally difference the series.[2] What happens if we take the latter route instead?

The SACF and SPACF for the first-differenced and seasonally differenced directory assistance pre-intervention series are shown in Figure 5.6. The spike at lag 12 in the SACF practically jumps off the page. Because the SPACF includes spikes at all three seasonal lags, a reasonable first guess is that there is a seasonal MA component as well. The estimates from this $(0,1,0)(0,1,1)_{12}$ model are presented in the middle portion of Table 5.1, and the estimated equation can be written as

$$\Delta\Delta_{S12}Calls_t = -0.78 + a_t - a_{t-12}$$

where a_t is our white noise error term. The bounds of invertibility for an MA(1) process is $-1 < \theta < 1$, and our estimated $\theta_{12} = -1$. It therefore violates the bounds of invertibility. Such a series has an infinite variance, in expectation; in our particular realization, the estimated standard error of θ_{12} is 456 – another clue that the estimation has "blown up." This model does not pass the first set of diagnostic criteria, and so there is no point in examining the model residuals.

We still need to address the large estimated autocorrelation at lag 12 in the first- and seasonally-differenced series, though. What happens if we try a seasonal

2 Trying a seasonal MA(1) term is not unreasonable either, but will be left as an exercise.

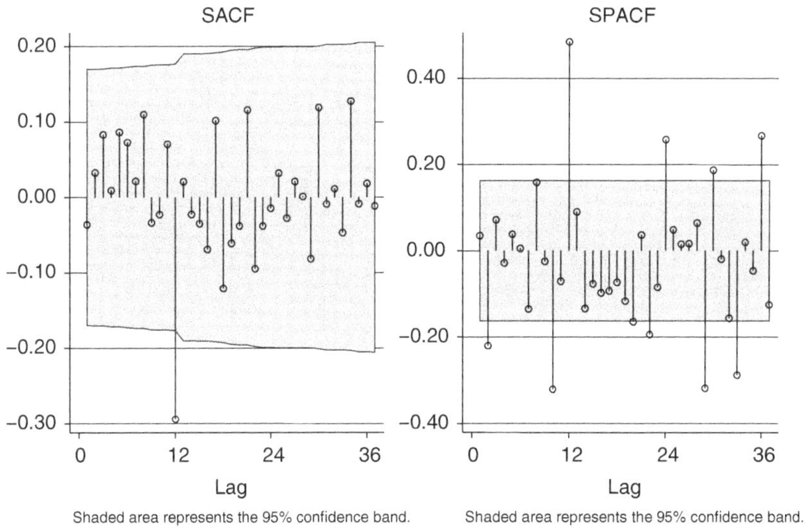

Figure 5.6 The SACF and SPACF for the first-differenced and seasonally-differenced pre-intervention series.

AR(1) component? The estimates from this $(0,1,0)(1,1,0)_{12}$ model appear in the last section of Table 5.1, and can be written as follows:

$$\Delta\Delta_{S12}Calls_t = -1.46 - 0.44\ \Delta\Delta_{S12}Calls_{t-12} + a_t,$$

where a_t is our white noise error term. The estimated ϕ_{12} is statistically significant, $p < 0.001$ and within the bounds of stationarity.

The SACF for the estimated residuals from this model are shown in Figure 5.7. There are no significant spikes anywhere in the SACF. The associated Q-tests through lag 24 are presented in Table 5.3.

No p-value falls below even .10; we do not find evidence of autocorrelation in the residuals. This model, then, is also an adequate model.

If we were at the final stage of our analysis, we would then proceed to the "meta-diagnosis" stage; we would compare the information criteria associated with the two models, and choose the model with the lowest values. The AIC and SBC for both models are presented in the rightmost two columns of Table 5.1. In this case, the last model – $(0,1,0)(1,1,0)_{12}$ would be preferred because its AIC and SBC of 1255 and 1264, respectively, are lower than those of the $(0,1,0)(1,0,0)_{12}$ model (1361 and 1370, respectively). Our task in Chapter 6, though, will be to estimate the best model for the entire series, including a component modelling the effect of the intervention. We will keep both models as candidates for this effort.

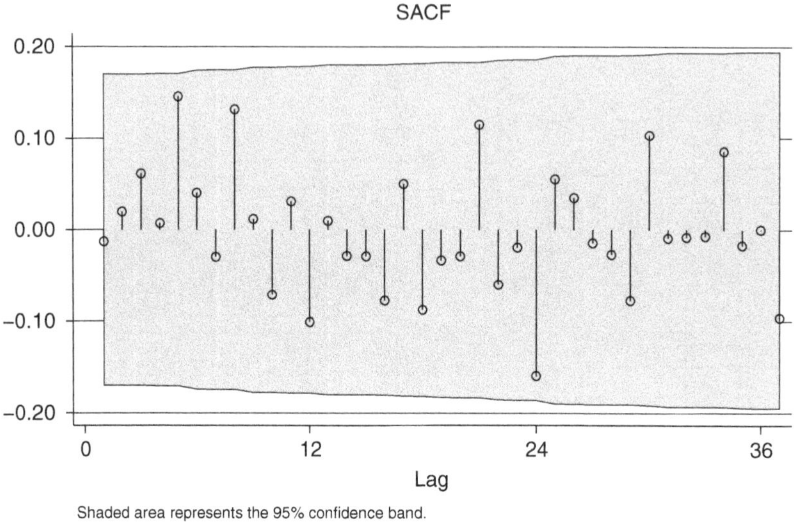

Figure 5.7 The SACF and SPACF for the $(010)(110)_{12}$ series.

Table 5.3 ARIMA $(0,1,0)(1,1,0)_{12}$ model Q-tests.

Lag	*Q*	*P*-value	Lag	*Q*	*P*-value
1	0.02	0.88	13	8.8	0.79
2	0.08	0.96	14	9.0	0.83
3	0.59	0.90	15	9.1	0.87
4	0.60	0.96	16	10.0	0.87
5	3.58	0.61	17	10.4	0.89
6	3.81	0.70	18	11.6	0.87
7	3.93	0.79	19	11.8	0.90
8	6.41	0.60	20	11.9	0.92
9	6.43	0.70	21	14.0	0.87
10	7.17	0.71	22	14.6	0.88
11	7.31	0.77	23	14.7	0.91
12	8.82	0.72	24	18.8	0.76

5.4 Forecasting

Because vector autoregression (VAR) forecasts outperform ARIMA forecasts, especially for longer-term predictions, there is rarely a good reason to use ARIMA forecasts for research purposes. They are also a standard topic in time series texts; McCleary and Hay (1980:204–225) provide especially readable coverage. The topic is presented here only for completeness, and to illustrate four concepts. First, forecasts have two pieces: the forecast point estimate and the confidence interval around it. Second, the point estimate can be calculated in either of two ways: the conditional forecast uses the preferred ARIMA model and observed values in the series, while the unconditional forecast uses only the model. Third, the forecast profile – i.e., the time path of the forecasts and the intervals around them – differ for AR and MA processes. Finally, just like predicted values in ordinary regression, there are two sources of variability around the point forecast: variability due to the variance of the random shocks a_{t+1}–a_{t+k}, and variability due to the variance of the model estimates.

5.4.1 Point Estimates: Unconditional and Conditional Forecasts

The first step in generating forecasts will be to estimate an ARIMA model using the available data. We then assume that the model holds through the period we are interested in forecasting, and extrapolate. If the assumption of model constancy is incorrect, of course, we should not have much faith in forecasts from it. We will find it useful to have not only good forecasts, but also some indication of the precision of the estimate. If the confidence interval around our forecast is large, after all, an unbiased point estimate for the forecast may not be especially helpful.

Let $Y_t(n)$ denote a forecast of Y_{t+n} made at time t. For example, $Y_t(1)$ denotes the forecast for the period $t+1$.

The point estimate can be calculated in either of the following two ways.

- The *conditional forecast*, $Y_t^c(n)$, is defined as $E(Y_t(n) \mid Y_1, Y_2 \ldots Y_t)$ – i.e., it includes the information about the specific realized values that were used to develop the model.
- The *unconditional forecast*, $Y_t^u(n)$, is defined as $E(Y_t(n) \mid model)$. It is based only on the estimated model, and does not use the data we have from previous periods (except for estimating the model).

To illustrate, consider the following AR(1) process. To simplify calculations, and without loss of generality, we will assume a constant term of 0.

$$Y_t = \phi Y_{t-1} + a_t.$$

It is simple enough to write

$$Y_{t+1} = \phi Y_t + a_{t+1}$$

and take expectations:

$$\begin{aligned} E(Y_{t+1}) = E(\phi Y_t) + E(a_{t+1}) \\ = \phi Y_t + 0 \\ = \phi Y_t. \end{aligned}$$

This expression, which uses the realized value Y_t, gives us the conditional forecast $Y_t^c(1)$.

For a two-step-ahead forecast,

$$\begin{aligned} Y_t^c(2) &= E(\phi^2 Y_t + \phi a_{t+1} + a_{t+2}) \\ &= E(\phi^2 Y_t) + \phi E(a_{t+1}) + E(a_{t+2}) \\ &= \phi^2 Y_t. \end{aligned}$$

More generally,

$$Y_t^c(n) = \phi^n Y_t.$$

The unconditional forecast, however, uses the ARIMA model but not the realized values Y_1–Y_t. It can be derived by expressing the series in its MA form, and then taking expectations.

$$\begin{aligned} Y_t &= \phi Y_{t-1} + a_t \\ (1-\phi L)Y_t &= a_t \\ Y_t &= \frac{1}{1-\phi L} a_t \\ &= (1 + \phi L + \phi^2 L^2 + \phi^3 L^3 + \cdots) a_t \\ &= a_t + \phi a_{t-1} + \phi^2 a_{t-2} + \phi^3 a_{t-3} + \cdots. \end{aligned}$$

Taking expectations,

$$\begin{aligned} Y_t^u(n) &= E(a_{t+1}) + \phi E(a_t) + \phi^2 E(a_{t-1}) + \cdots \\ &= 0 + 0 + 0 + \cdots \\ &= 0. \end{aligned}$$

That is, $E(Y_t(n)) = 0$.

Notice that for the AR(1) series, the conditional forecast changed over time, converging to the series mean; the unconditional forecast, however, is constant over time and equal to the series mean.

The root mean-square forecast error (RMSFE) can be used to summarize the deviation of forecasts from their realized values.

$$RMSFE = 1/n\sqrt{\sum_{i=1}^{n}[Y_{t+i} - Y_t(i)]^2}.$$

It can be calculated, of course, only after we have observations for the forecast period. It is easily shown that conditional forecasts have a lower MSFE than unconditional forecasts. It should not be surprising that the former outperform the latter since the conditional forecast uses more information.

5.4.2 Forecast Errors for Interval Estimates

Finally, we need to derive forecast errors in order to construct confidence intervals around our forecasts. The forecast error e_{t+n} is defined as the difference between the realized value and the forecast value:

$$e_{t+n} = Y_{t+n} - Y_{t(n)}$$

for both conditional and unconditional forecasts. To derive a general form, recall from Wold's Theorem that any weakly stationary stochastic process $Y_t - \mu$, where μ is the series mean, can be written in terms of weighted moving average components, as in Equation (4.3). For the purposes of our derivation, it will be convenient to rewrite this equation in ψ-weight form as follows:

$$Y_t - \mu = a_t + \psi_1 a_{t-1} + \psi_2 a_{t-2} + \ldots + \psi_k a_{t-k}. \tag{5.4}$$

Then

$$Y_{t+1} - \mu = a_{t+1} + \psi_1 a_t + \psi_2 a_{t-1} + \psi_3 a_{t-2} + \cdots + \psi_k a_{t-k+1} + \cdots. \tag{5.5}$$

5.4.2.1 The Forecast Error When the True ARIMA Model is Known

We can write the forecast error, e_{t+n} as the difference between the realized Y_{t+n} and our forecast, $Y_t^c(n)$.

Then the general form for the forecast error for a one-step-ahead forecast becomes

$$\begin{aligned}
e_{t+1} &= Y_{t+1} - E[Y_{t+1} \mid Y_t, Y_{t-1}, Y_{t-2} \ldots] \\
&= Y_{t+1} - Y_t^c(1) \\
&= [a_{t+1} + \psi_1 a_t + \psi_2 a_{t-1} + \psi_3 a_{t-2} \\
&\qquad + \cdots + \psi_k a_{t-k+1} + \cdots] - [E(Y_{t+1} \mid Y_t, Y_{t-1}, Y_{t-2}, \ldots)] \\
&= [a_{t+1} + \psi_1 a_t + \psi_2 a_{t-1} + \psi_3 a_{t-2} + \cdots + \psi_k a_{t-k+1} + \cdots] \\
&\qquad - (E[a_{t+1}] + \psi_1 a_t + \psi_2 a_{t-1} + \psi_3 a_{t-2} + \cdots + \psi_k a_{t-k+1} + \cdots) \\
&= a_{t+1}
\end{aligned}$$

Then the variance of $Y_t^c(1)$, the conditional one-step-ahead forecast, is

$$Var[Y_t^c(1)] = Ee_{t+1}^2 = \sigma_a^2$$

For a two-step-ahead forecast,

$$Y_t^c(2) = EY_t^c(2) \mid Y_t, Y_{t-1}, Y_{t-2}, \ldots$$

i.e., we can incorporate all information except for a_{t+1} and a_{t+2}, since those values have not yet been observed. Then

$$\begin{aligned}
e_{t+2} &= Y_{t+2} - Y_t(2) \\
&= a_{t+2} + \psi_1 a_{t+1}
\end{aligned}$$

and so

$$\begin{aligned}
Var(2) &= Ee_{t+2}^2 \\
&= E[(a_{t+2} + \psi a_{t+1})^2] \\
&= Ea_{t+2}^2 + 2\psi_1 E(a_{t+2} a_{t+1}) + \psi_1^2 Ea_{t+1}^2 \\
&= \sigma^2 + \psi_1^2 \sigma^2 \\
&= (1 + \psi_1^2)\sigma^2
\end{aligned}$$

In similar fashion,

$$Var(3) = (1 + \psi_1^2 + \psi_2^2)\sigma^2$$

More generally,

$$Var(n) = (1 + \psi_1^2 + \psi_2^2 + \cdots + \psi_{n-1}^2)\sigma^2$$

As a consequence, forecast errors for pure MA processes are constant over time, while the forecast errors for AR processes increase over time.

5.4.2.2 Aside: Rewriting an ARMA Series as a Pure MA Process

At first glance, the connection between the ψ-weights and the estimated coefficients from our ARIMA model may be unclear. To illustrate, suppose we have a $(1,0,0)(0,0,1)_4$ model; ϕ is the AR coefficient, θ is the coefficient for the seasonal MA term, and to simplify the algebra (and with no loss of generality), we will omit a constant term. We can use lag operator algebra to rewrite the seasonal ARMA model as its MA representation (*psi*-weight form) as follows.

$$
\begin{aligned}
Y_t &= \phi Y_{t-1} + a_t - \theta a_{t-4} \\
(1-\phi L)Y_t &= (1-\theta L^4)a_t \\
Y_t &= \frac{1-\theta L^4}{1-\phi L} a_t \\
&= [(1-\theta L^4)(1+\phi L + \phi^2 L^2 + \phi^3 L^3 + \ldots)]a_t \\
&= [(1+\phi L + \phi^2 L^2 + \phi^3 L^3 + \phi^4 L^4 + \phi^5 L^5 + \ldots + \phi^k L^k + \ldots) \\
&\quad -(\theta L^4 + \phi\theta L^5 + \phi^2\theta L^6 + \ldots + \phi^{k-4}\theta L^k + \ldots)]a_t \\
&= [1+\phi L + \phi^2 L^2 + \phi^3 L^3 + (\phi^4 - \theta)L^4 + (\phi^5 - \phi\theta)L^5 + (\phi^6 - \phi^2\theta)L^6 \\
&\quad + \ldots + (\phi^k - \phi^{k-4}\theta)L^k + \ldots]a_t \\
&= a_t + \phi a_{t-1} + \phi^2 a_{t-2} + \phi^3 a_{t-3} + (\phi^4 - \theta)a_{t-4} + (\phi^5 - \phi\theta)a_{t-5} \\
&\quad + (\phi^6 - \phi^2\theta)a_{t-6} + \ldots + (\phi^k - \phi^{k-4}\theta)a_{t-k} + \ldots].
\end{aligned}
$$

Then for the $(1,0,0)(0,0,1)_4$ model, the first few ψ-weights are

$$
\begin{aligned}
\psi_1 &= \phi \\
\psi_2 &= \phi^2 \\
\psi_3 &= \phi^3 \\
\psi_4 &= \phi^4 - \theta \\
\psi_5 &= \phi^5 - \phi\theta \\
\psi_6 &= \phi^6 - \phi^2\theta.
\end{aligned}
$$

Remember that for ϕ to satisfy the bounds of stationarity and θ to satisfy the bounds of invertibility, $-1 < \phi < 1$ and $-1 < \theta < 1$. Then the ψ weights (or coefficients on the a_{t-k}) become infinitesimally small as k grows larger. As a practical matter, then, we will need only the first few ψ-weights in our forecast error calculations.

5.4.2.3 The Forecast Error When the True ARIMA Model is Unknown

In practice, of course, we estimate an ARIMA model from our data but the true DGP is unknown. In the presentation above, we derived the *psi*-weights as functions of our estimated ARIMA parameters ϕ and θ without incorporating the variance around these estimates. The true forecast error includes those variances, and so precise estimates of the ARIMA coefficients are critical for obtaining tight confidence intervals around our forecasts.

5.5 Summary

- ARIMA estimation is useful for clarifying the nature of the autocorrelation in our series even if our long-run goal is not ARIMA modeling.
- ARIMA$(p,d,q)(P,D,Q)_S$ notation provides convenient shorthand for ARIMA model specification. The p, d, and q indicate the regular AR-order, degree of regular differencing and regular MA-order, respectively. The P, D, and Q indicate the seasonal AR-order, degree of seasonal differencing, and seasonal MA-order. S represents the number of time points in a season.
- The ARIMA model-building process involves several steps: identification, estimation, diagnosis, and meta-diagnosis. Specific criteria must be met before advancing from one step to the next. An adequate ARIMA model has no remaining autocorrelation in its residuals.
- Because vector autoregression forecasts outperform ARIMA forecasts, ARIMA forecasts are no longer widely-used for research purposes.

5.6 Appendix to Chapter 5: Non-linear Models and Numerical Estimation Methods

To see why ordinary least squares (OLS) cannot be used to estimate ARIMA models, first recall a standard regression model

$$Y_i = \beta_0 + \beta_1 X_i + e_i.$$

For simplicity's sake, use the mean-deviation form:

$$y_i = Y_i - \bar{Y},\ x_i = X_i - \bar{X}$$

$$y_i = \beta_1 x_i + e_i.$$

OLS involves minimizing the sum of the squared error terms:

$$e_i = y_i - \beta_1 X_i$$

$$\sum_{i=1}^{n} e_i^2 = \sum_{i=1}^{n} y_i^2 - 2\beta_1 \sum_{i=1}^{n} x_i y_i + \beta^2 \sum_{i=1}^{n} x_i^2.$$

If we take the derivative of this, and set it equal to zero (to find the minimum), we get a "least-squares normal equation" that is *linear* in the parameter β, and therefore easily solved:

$$\frac{\partial SSE}{\partial \beta} = -2\sum x_i y_i + 2\hat{\beta}\sum x_i^2 = 0$$

$$\hat{\beta} = \frac{\sum x_i y_i}{\sum x_i^2}$$

Unfortunately, we cannot derive a simple equation for an OLS estimator for ARIMA models.

To see this, consider an MA(1) model,

$$Y = a_t - \theta a_{t-1}.$$

To minimize the sum of the squared errors, rewrite this as

$$\begin{aligned} Y_t &= a_t - \theta L a_t \\ &= (1-\theta L)a_t \end{aligned}$$

or

$$\begin{aligned} a_t &= \frac{1}{1-\theta L} Y_t \\ &= (1+\theta L + \theta^2 L^2 + \theta^3 L^3 + \cdots)Y_t \\ &= Y_t + \theta Y_{t-1} + \theta^2 Y_{t-2} + \theta^3 Y_{t-3} + \cdots \\ &= \sum_{k=0}^{\infty} \theta^k Y_{t-k}. \end{aligned}$$

Then the sum of the squared errors/shocks becomes

$$\sum_{t=1}^{n} a_t^2 = \sum_{t=1}^{n} \left[\sum_{k=0}^{\infty} \theta^k Y_{t-k} \right]^2.$$

Taking the derivative of this and setting it to 0, we get

$$\begin{aligned} \frac{\partial \text{SSE}}{\partial \theta} &= 2\sum_{t=1}^{n} \left(\sum_{k=0}^{\infty} \theta^k Y_{t-k} \right) \left(k \sum_{k=0}^{\infty} \theta^{k-1} Y_{t-k} \right) = 0 \\ &= 2\sum_{t=1}^{n} (Y_t + \theta Y_{t-1} + \theta^2 Y_{t-2} + \cdots)[k(1/\theta Y_t + Y_{t-1} + \theta Y_{t-2} + \theta^2 Y_{t-3} + \cdots)] = 0. \end{aligned}$$

This normal equation, which is non-linear in the parameter of interest (θ), cannot be solved analytically. Instead, we will have to use a "numerical estimation" method, such as maximum likelihood estimation. In the case of our MA(1) model, this process would entail the following.

1) Choose a starting value for θ.
2) Generate predicted values for Y_t based on this initial guess.
3) Compare the predicted values to the real values, using some criterion.
4) Choose a different value for θ.
5) Generate predicted values for Y_t based on this *new* guess.
6) Compare these predicted values to the real values.
7) If these predicted values are closer to the real values than the first set of predicted values were, try another θ close to this second guessed θ.
8) Otherwise, if the first set of predictions was better than the second, try another θ closer to the first guessed θ.

9) Keep doing this until you are getting predictions that are "close together" by some standard. (Sometimes, e.g., second or third decimal.).

These days, modern software with sound ARIMA estimation procedures automates this process and chooses promising starting values. In the early days of ARIMA modelling, though, researchers needed to provide their own starting values. The cover art for this book includes a graphic provided by Box and Jenkins (1970:519) for obtaining starting values for MA(2) models. The arcs inside the the pyramid represent values of the sample autocorrelation at lags 1 and 2. Using the point of the intersecting SACF(1) and SACF(2) arcs, the starting value for θ_1 is taken from the x-value and the starting value for θ_2 is taken from the y-value.

References

Box, G.E.P. and Jenkins G.M. (1970). Time series analysis: forecasting and control San Francisco: Holden-Day.

McCleary R. and Hay R.A., Jr. (1980). Applied Time Series Analysis for the Social Sciences Beverly Hills, CA: Sage.

McSweeny, A.J. (1978). The effects of response cost on the behavior of a million persons: charging for directory assistance in Cincinnati. *Journal of Applied Behavioral Analysis* 11:47–51.

6

ARIMA Intervention Models

Chapter 1 showcases the difficulty behind deciding if the 2007 US troop surge in Iraq led to a decrease in US troop fatalities. While the surge was followed by a sharp drop in the number of deaths, it is unclear whether the drop was associated with the surge or merely part of the normal fluctuation in the series. This is an example of a problem that may be addressed in an ARIMA context if necessary assumptions are met: we can model the baseline variation as an ARIMA process and then test whether there is a deviation from the baseline that is associated with the intervention. This is the strategy underpinning "ARIMA impact assessment" associated with a class of models generally referred to as "ARIMA intervention models." Modeling the impacts of interventions this way is especially useful when we either do not know or cannot measure key drivers of the baseline variation.

These models are based on three critical assumptions. First, we must know when the event occurs and have sufficient information to model it. Typically, the event or intervention will be modeled as an off/on (0/1) indicator variable; the timing must be known in advance. Second, the intervention is assumed to affect the dependent variable, but the dependent variable must not affect the intervention. Vector autoregression (VAR) models, covered in Chapter 12, are able to address situations in which this assumption cannot be met. (Indeed, since ARIMA models can be written as special cases of VAR models, it is common practice today to begin with VAR models and test restrictions to decide whether the special cases are appropriate.) Finally, the baseline modeled in ARIMA terms is assumed to be the same throughout the period – for example not affected by (or even spuriously related to) the intervention. Then the baseline and intervention components can be treated as additive, and we can write

$$Y_t = \upsilon(I_t) + N_t, \tag{6.1}$$

Applied Time Series Analysis for the Social Sciences: Specification, Estimation, and Inference, First Edition. Regina M. Baker.

Companion website: www.wiley.com/go/baker

where

- I_t is the intervention,
- $\upsilon(I_t)$ is a *transfer function* used to model the response of the series to the intervention as time passes, and
- N_t is the ARIMA "noise" component, our baseline.

The modeling strategy is based on the assumption that if the counterfactual were true – if there were no intervention – then the series would continue to follow a path consistent with our ARIMA model; any significant deviation is presumed to be an effect of the intervention.

The intervention I_t is typically modeled as either a pulse function – that is, a series such as "0. . .0,0,0,0,1,0,0,0. . .0" – or a step function such as "0. . .0,0,0,0,0,1, 1,1,1,1,. . .1," shown in Figure 6.1a and 6.1b, respectively. When working with a raw series, pulse functions represent events that occur in a single period, such as an important news announcement or a single bombing. Step functions represent long-term shifts, such as a new tax or a new law.

This chapter will begin by introducing the general form of transfer functions. In practice, however, most responses to interventions can be reasonably approximated by simpler forms, "zero-order" transfer functions and "first-order" transfer functions covered in sections 6.2 and 6.3, respectively. Zero-order transfer functions are appropriate when modeling a one-time change in the series or a permanent change to a stable new level. "First-order transfer functions" are used to model certain types of gradual changes. Understanding why these transfer function models are associated with these specific responses in the dependent variable is critical to understanding dynamic models in general and will reappear in Chapters 9–12 covering autoregressive distributed lag models and VAR approaches. We will close this chapter by estimating a zero-order transfer function model of the response to the charge for directory assistance (411) calls in the series introduced in Chapter 5.

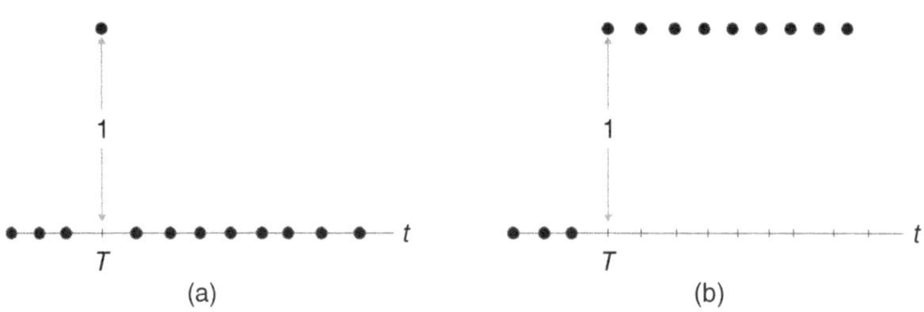

Figure 6.1 Common models for the intervention, I_t. (a) A pulse function and (b) a step function.

6.1 Transfer Functions: General Form

A transfer function $\upsilon(I_t)$ models the response of the series, over time, to the intervention. It can take any number of forms, permitting us to model a rich set of possibilities. The sequelae are collectively referred to as the *impulse response* and often illustrated graphically. Some examples of transfer functions and impulse response functions associated with a pulse intervention – for example, a bit of news or an event – are provided in Figure 6.2.

We see the simplest possible response in Figure 6.2a. The transfer function is written as $\upsilon(I_t) = \omega_0 I_t$, indicating that we expect the series to deviate from our ARIMA baseline model by $\omega_0 I_t$. Since our pulse function I_t is 0 for all t except the single period of the intervention, T, we expect no deviation from baseline before or after time T. At time T, however, $I_t = 1$, so we expect a deviation from our ARIMA baseline of ω_0. To illustrate, suppose we are interested in the response of alumnae donations to a university that just won a high-profile football game. In this case, we have modeled it as if we expect donations to increase by the amount ω_0 in the time period that included the victory but to return to ordinary levels immediately afterward.

Effects of the intervention need not be confined to the same period, though. Figure 6.2b shows a transfer function and impulse response for a case where the intervention has an effect both in the contemporaneous period and in the period

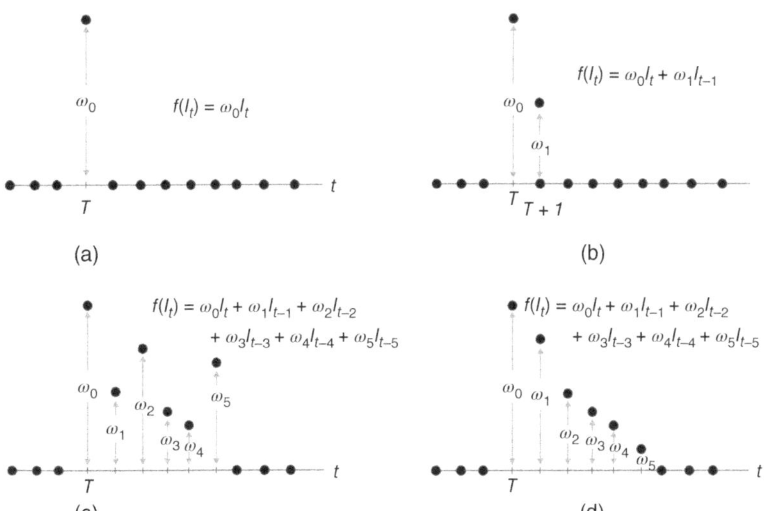

Figure 6.2 Some possible transfer functions and their impulse response profiles when I_t is a pulse function.

immediately following. At T, the period of the intervention, we expect the deviation from the ARIMA baseline to be ω_0; at time $T+1$, the period immediately following the intervention, we expect a deviation from the baseline to be ω_1.

In theory, a transfer function and its associated impulse response function could be as messy as the example shown in Figure 6.2c. The largest effect of the intervention is observed in the contemporaneous period; the effect in the next period is about half as large. The effect of the intervention in the second period after the intervention is larger than that in the first period ($\omega_2 > \omega_1$), but the effects in the third and fourth periods after the intervention are smaller than those in the first period. It is difficult to imagine an intervention that would have such an effect. In real life, it is more common for the effects of an intervention to decrease as time passes, as in Figure 6.2d. Moreover, in a case such as Figure 6.2d, we are often unable to estimate $\omega_0 - \omega_5$ precisely with the number of time points we have. Fortunately, most of the impulse responses we are likely to encounter are well approximated by two general forms, zero-order transfer functions and first-order transfer functions, and most of the remaining impulse responses are well approximated by a combination of these simpler forms.

The general form of a transfer function can be written as

$$\upsilon(I_t) = \frac{\omega(L)L^l}{\delta(L)}, \tag{6.2}$$

where

$$\omega(L) = \omega_0 - \omega_1 L - \omega_2 L^2 - \cdots - \omega_s L^s \tag{6.3}$$

and

$$\delta(L) = 1 - \delta_1 L - \delta_2 L^2 - \cdots - \delta_r L^r, \tag{6.4}$$

and l indicates the timing of the onset of the response. $l = 0$, for example, indicates that the response begins in the same period as the event or initial shift. $l = 1$ indicates that the response begins in the first period after the initial shift; $l = 2$ indicates that the response begins in the second period after the response, etc.

To see how the shape of the impulse response function is affected by different parameters ω, δ, s, and r, it will be helpful to examine a couple of simple forms, "zero-order" transfer functions and "first-order" transfer functions, defined by the value of r. If the denominator of the transfer function does not include lagged terms, the function is a zero-order transfer function and all responses will occur only in the periods included in the numerator. A first-order transfer function includes $1-\delta L$ in the denominator and is used to model gradual effects. Illustrations are provided in Sections 6.2 and 6.3.

6.2 Simplest Form: Zero-Order Transfer Functions

Remember that in order to model the noise (stochastic) portion of the series in ARIMA terms, we must have a stationary series. Transfer functions, represent the pattern of the deviation of the series mean that results from the intervention.

6.2.1 Form and Notation

The simplest form of response to an intervention is a shift in the mean of the series by a constant amount, as in Figure 6.3a and 6.3b. The transfer function $\upsilon(I_t)$ can be written

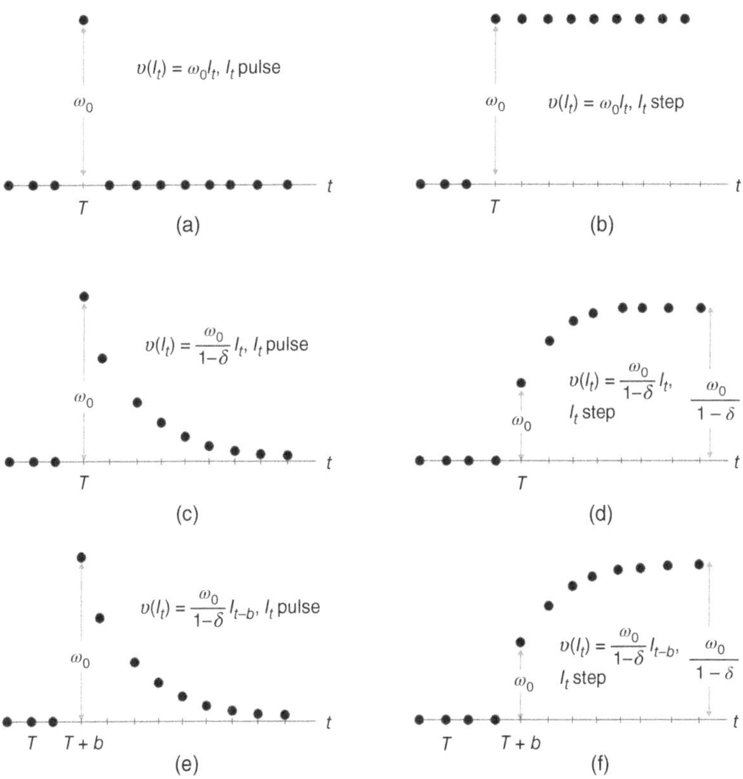

Figure 6.3 Responses associated with commonly-used transfer functions.

$$\upsilon(I_t) = \omega_0 I_t. \tag{6.5}$$

Then before the intervention,

$$\begin{aligned}\upsilon(I_t) &= \omega_0 I_t \\ &= \omega_0 \times 0 \\ \upsilon(I_t) &= 0\end{aligned}$$

That is, the series is fully represented by only the noise model (the ARIMA portion).

6.2.2 A Zero-order Transfer Function with a Pulse Intervention

Now suppose the intervention is a one-period event – that is best represented by a pulse function – and let T be the period of the intervention.

Then at $t = T$,

$$\begin{aligned}\upsilon(I_t) &= \omega_0 I_t \\ &= \omega_0 \times 1 \\ \upsilon(I_t) &= \omega_0.\end{aligned} \tag{6.6}$$

That is, during the single period of the intervention, the series is shifted by ω_0.

After that period, $I_t = 0$ and so $\upsilon(I_t) = 0$, as in Equation (6.3a). To summarize, the response of Y_t to the intervention is a one-period shift in the mean of the series, as shown in Figure 6.3a.

6.2.3 A Zero-order Transfer Function with a Step Intervention

If the intervention I_t is a step function, the pre-intervention period will again reduce to the noise model, since $\omega_0 \times 0 = 0$. From the first period of the intervention on, though, $f(1) = \omega_0$. That is, the response will appear to be a permanent shift in the series mean by ω_0, as in Figure 6.3b.

6.3 Gradual Changes: First-Order Transfer Functions

6.3.1 Form and Notation

In some cases, the full effect of the intervention occurs over several periods. Two common patterns that can be modeled with a simple first-order transfer function are shown in Figure 6.3c and 6.3d. The transfer function is the same for both. When applied to a pulse function, though, the response is a change (by ω_0) in the intervention period, followed by a gradual return to the series' original level. When applied to a step function, the response is a change by ω_0 in the first period of the intervention, followed by a gradual shift to a new level that differs from the previous level by $\frac{\omega_0}{1-\delta}$. To see why this is the case, let us take a closer look at the first-order transfer function:

$$\upsilon(I_t) = \frac{\omega_0}{1-\delta L} I_t, \tag{6.7}$$

where $-1 < \delta < 1$.

Recall that since the intervention model is additive, we can examine the behavior of the transfer function separately from the behavior of the noise (N_t, or ARIMA) portion of the model:

$$Y_t = \upsilon(I_t) + N_t. \tag{6.8}$$

Then let

$$\begin{aligned} Y_t^* &= Y_t - N_t \\ &= \upsilon(I_t) \\ &= \frac{\omega_0}{1-\delta L} I_t \\ (1-\delta L)Y_t^* &= \omega_0 I_t \\ Y_t^* &= \delta Y_{t-1}^* + \omega_0 I_t. \end{aligned}$$

6.3.2 First-Order Transfer Functions with Pulse Interventions

Let T be the time of the pulse; that is

$$I_t = \begin{cases} 0, & \text{for all} t \neq T, \\ 1, & \text{if} t = T. \end{cases}$$

In the preintervention period, $I_t = 0$. Then it is clear from Equation (6.7) that for all $t < T, Y_t^* = 0$.

Now consider what happens at $t = T$, the intervention period.

$$\begin{aligned} Y_T^* &= \delta Y_{T-1}^* + \omega_0 I_t \\ &= 0 + \omega_0 \times 1 \\ &= \omega_0. \end{aligned}$$

In the period immediately following the intervention period,

$$\begin{aligned} Y_{T+1}^* &= \delta Y_T^* + \omega_0 I_t \\ &= \delta\omega_0 + \omega_0 \times 0 \\ &= \delta\omega_0. \end{aligned}$$

In the next period, $t = T + 2$,

$$\begin{aligned} Y_{T+2}^* &= \delta Y_{T+1}^* + \omega_0 I_t \\ &= \delta(\delta\omega_0 + \omega_0 \times 0) \\ &= \delta^2 \omega_0. \end{aligned}$$

The pattern should be becoming clear. At period $t = T + k$, $k \geq 0$,

$$Y_{T+k}^* = \delta^k \omega_0. \tag{6.9}$$

Since $-1 < \delta < 1$ – i.e. δ is a fraction – δ^k becomes smaller in absolute value as k increases: as we move further past the intervention period T. Indeed,

$$\lim_{k \to \infty}(\delta^k \omega_0) = 0.$$

In the long run, the response of the dependent variable to the first-order transfer function applied to a pulse function is to return asymptotically to the preintervention level of the series, as illustrated in Figure 6.4a.

6.3.3 First-Order Transfer Functions with Step Interventions

If our intervention is a step function rather than a pulse function, the first-order transfer function will result in a gradual change to a new level, as in Figure 6.4b.

Once more, consider only the transfer function portion (Y_t^*) of the ARIMA intervention model:

$$\begin{aligned} Y_t^* &= Y_t - N_t \\ &= \frac{\omega_0}{1 - \delta L} I_t. \end{aligned}$$

In this case, let T represent the *first period* of the intervention. During the pre-intervention portion of the series, $t < T$, $I_t = 0$, so

$$Y_t^* = \frac{\omega_0}{1 - \delta L} I_t = 0.$$

At the onset of the intervention, $t = T$,

$$\begin{aligned} Y_T^* &= \frac{\omega_0}{1 - \delta L} I_t \\ (1 - \delta L) Y_T^* &= \omega_0 I_t \\ Y_T^* &= \delta Y_{T-1}^* + \omega_0 I_t \\ &= \delta(0) + \omega_0(1) \\ &= \omega_0. \end{aligned}$$

At the second period of the intervention, $t = T + 1$,

$$Y_{T+1}^* = \frac{\omega_0}{1 - \delta L} I_t$$

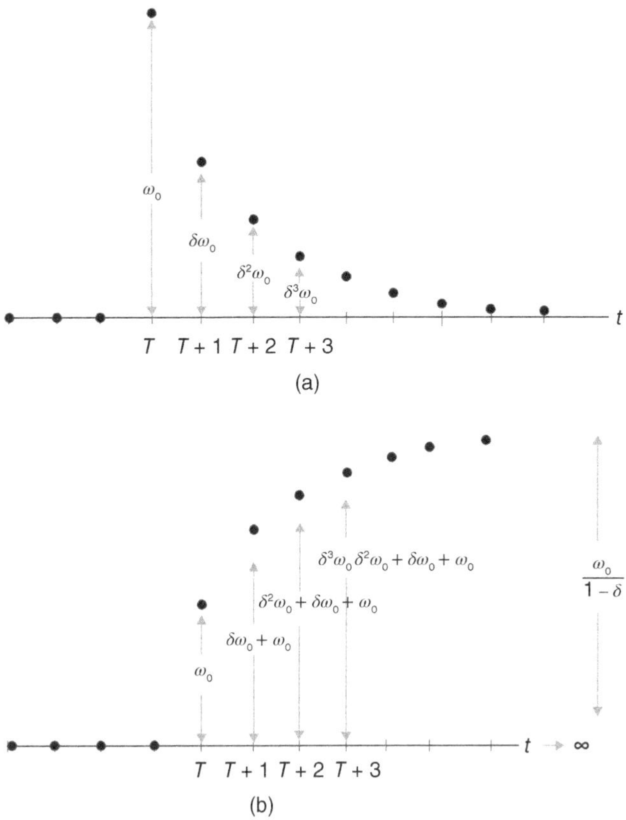

Figure 6.4 Common models for the intervention, I_t. (a) A pulse function and (b) a step function.

$$\begin{aligned}(1-\delta L)Y^*_{T+1} &= \omega_0 I_t\\ Y^*_{T+1} &= \delta Y^*_T + \omega_0 I_t\\ &= \delta\omega_0 + \omega_0.\end{aligned}$$

At the third period of the intervention, $t = T + 2$,

$$\begin{aligned}Y^*_{T+2} &= \frac{\omega_0}{1-\delta L} I_t\\ (1-\delta L)Y^*_{T+2} &= \omega_0 I_t\\ Y^*_{T+2} &= \delta Y^*_{T+1} + \omega_0 I_t\\ &= \delta[\delta\omega_0 + \omega_0] + \omega_0(1)\\ &= \delta^2\omega_0 + \delta\omega_0 + \omega_0.\end{aligned}$$

In general, for the nth (post)intervention case,

$$Y^*_{T+n} = \sum_{n}^{k=0} \delta^k \omega_0.$$

Notice that the permanent change in the level of the series (as $n \to \infty$) becomes

$$\omega_0 + \delta\omega_0 + \delta^2\omega_0 + \delta^3\omega_0 + \cdots = \frac{1}{1-\delta}\omega_0$$
$$= \frac{\omega_0}{1-\delta}.$$

6.4 Example: The Directory Assistance (411) Series

In Chapter 5, we examined the number of calls to directory assistance (411) in Cincinnati before and after the phone company began charging for them. Two features of Figure 5.3, the plot of the raw series, were striking. First, the series is nonstationary; the number of calls increased dramatically through the pre-intervention period, and after a dramatic drop when the charge was introduced, increased again. Second, there was a sharp drop in the number of calls when the charge was introduced. The effect occurred quickly enough so that we will follow McCleary and Hay's example and model it as an immediate drop.

The first decision we must make is how to model the intervention. In the case of this non-stationary series, there are two ways to think about I_t. First, let us think about the logic of the problem: there was no charge for 411 calls in the pre-intervention period, but once the charge began, it continued through the rest of the series. This suggests a step function: $I_t = 0$ in the pre-intervention portion and $I_t = 1$ once the charge began. The response – the number of calls – is step like, in the sense that the trend appears to be shifted by a constant amount once the charge began. Remember, though, that we needed to difference the series to achieve stationarity. To preserve the relationship between the two raw series (the step function and the calls series), we must difference the right-hand side (the intervention/ step function) before estimating the transfer function model.

To see this from another perspective, a second way of thinking about the problem is to examine the differenced series, shown in Figure 6.5. Notice that when we difference the series, its shape changes from a permanent level shift to a one-period level shift. That is, although the response in the raw series is step-like, the response in the differenced series is a pulse. Indeed, differencing a step function yields a pulse function, a result you will demonstrate in an exercise. The following representations of the transfer function portion of the series are mathematically equivalent:

$$\Delta Y_t^* = \omega_0 \Delta I_t,$$

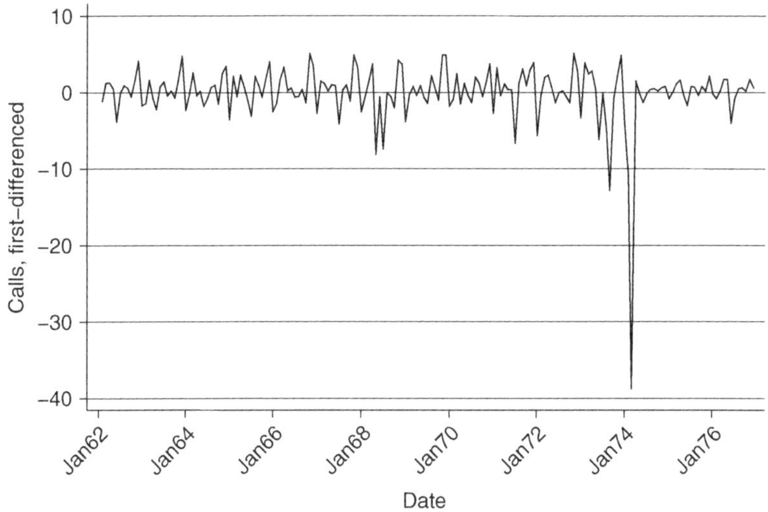

Figure 6.5 The directory assistance series after first differencing.

where I_t is a step function, and

$$\Delta Y_t^* = \omega_0 I_t,$$

where I_t is a pulse function.

Recall from Chapter 5 that our pre-intervention series was adequately represented by both a $(0,1,0)(1,0,0)_{12}$ and a $(0,1,0)(0,0,1)_{12}$ model, but we preferred the former both because the model residuals were cleaner because they had lower AIC and BIC. We will estimate both models with the transfer function.

The estimates are presented in Table 6.1. The estimated equation for the seasonal AR model can be written as follows:

$$\Delta \text{Calls}_t = 0.164 - 39.9\Delta I_t - 0.371\Delta \text{Calls}_{t-12} + a_t,$$

where I_t is a step function, $n = 179$, a_t is our white noise error term, and $\hat{\sigma}_a = 2.45$.

Recall our criteria for the "diagnosis" stage. First, the estimated coefficients on any AR and/or MA components should be statistically significant and must be within the bounds of stationarity and/or invertibility. $\phi_{12} = 0.371$ is statistically significant ($p < 0.001$), and within the bounds of stationarity for an AR(1) model: $-1 < \phi_{12} < 1$.

Second, the estimated residuals should appear to be white noise; that is, the SACF of the estimated residuals should not have spikes at key lags, and we should be able to reject the null hypothesis of joint statistical significance of autocorrelations using the Q-test.

Table 6.1 ARIMA estimates for the 411 series with a zero-order transfer function.

ARIMA model	Parameter	Estimate	Standard error	*P*-value	AIC	SBC
$(0,1,0)(1,0,0)_{12}$	θ_0	0.164	0.279	0.557	836	843
	ϕ_{12}	0.371	0.070	< 0.001		
	ω_0	−39.9	2.29	< 0.001		
$(0,1,0)(0,0,1)_{12}$	θ_0	0.234	0.234	0.497	843	852
	θ_{12}	−0.260	0.074	< 0.001		
	ω_0	−39.5	2.44	< 0.001		

P-values are reported for two-sided hypothesis tests of $\phi_k = 0, \theta_k = 0, \omega_0 = 0$.

The SACF for residuals from the $(0,1,0)(1,0,0)_{12}$ model are shown in Figure 6.6. The associated Q-tests are presented in Table 6.2.

In this case, there are large spikes at lags 22 and 24. The latter is noteworthy because it occurs at a seasonal lag; however, it is also important to note that there was no a statistically significant spike at lag 12. This suggests that we are not missing something that occurs annually but should think about whether or not we might be missing something that occurs every *two* years. At the same time,

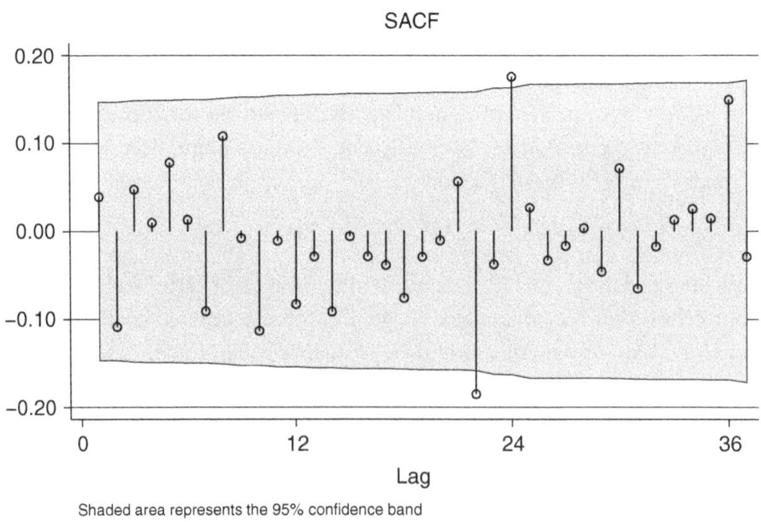

Figure 6.6 The SACF for the residuals of the $(0,1,0)(1,0,0)_{12}$ model with a zero-order transfer function.

Table 6.2 Q-tests for the residuals from the ARIMA $(0,1,0)(1,0,0)_{12}$ model with a zero-order transfer function.

Lag	Q	P-value	Lag	Q	P-value
1	0.28	0.60	13	11.8	0.55
2	2.43	0.30	14	13.4	0.50
3	2.85	0.42	15	13.4	0.57
4	2.87	0.58	16	13.6	0.63
5	4.00	0.55	17	13.9	0.68
6	4.03	0.67	18	15.0	0.66
7	5.58	0.59	19	15.2	0.71
8	7.80	0.45	20	15.2	0.76
9	7.81	0.55	21	15.9	0.78
10	10.3	0.42	22	22.9	0.41
11	10.3	0.51	23	23.2	0.45
12	11.6	0.48	24	29.7	0.20

though, the Q-test at lag 24 indicates that we are able to reject the null hypothesis that any of the autocorrelations through lag 24 are statistically significant and standard levels ($p = 0.20$). This model is therefore adequate. Modeling an important predictor that occurs every two years might improve it if we could identify one.

The estimates for the intervention model with an MA component are presented in the bottom half of Table 6.1. The associated equation can be written as

$$\Delta \text{Calls}_t = 0.234 - 39.5\Delta I_t + 0.260 a_{t-12} + a_t,$$

where I_t is a step function, $n = 179$, a_t is our white noise error term, and $\hat{\sigma}_a = 2.52$.

This model also passes the first set of criteria for the diagnosis stage: $\theta_{12} = -0.260$ is statistically significant ($p < 0.001$), and within the bounds of invertibility for an MA(1) model: $-1 \leq \theta_{12} \leq 1$. However, it does not pass the second set of criteria.

The SACF for the residuals from the $(0,1,0)(0,0,1)_{12}$ model are shown in Figure 6.7, and the associated Q-tests are presented in Table 6.3.

As with the seasonal AR model, we see spikes at lags 22 and 24, but not at lag 12. There are also spikes that extend outside the 95% confidence bounds at lags 2, 10, and 36, and based on the Q-test at lag 24, we are able to reject the null hypothesis that $\rho_k = 0$ for all k at the $p < .001$ level. This model is not adequate; it has failed at the diagnosis stage. There is no need to proceed to the metadiagnosis stage and compare information criteria. We choose the ARIMA $(0,1,0)(1,0,0)_{12}$ model. Recall

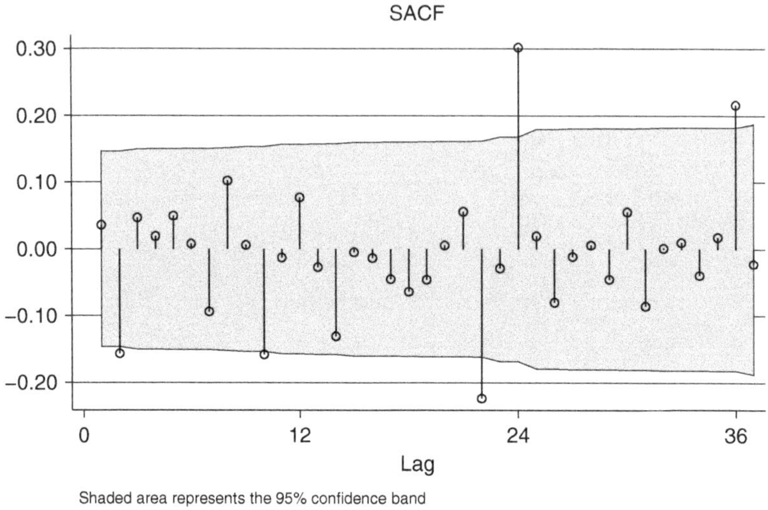

Figure 6.7 The SACF for the residuals of the $(0,1,0)(0,0,1)_{12}$ model with a zero-order transfer function.

Table 6.3 Q-tests for the residuals from the ARIMA $(0,1,0)(0,0,1)_{12}$ model with a zero-order transfer function.

Lag	Q	P-value	Lag	Q	P-value
1	0.235	0.63	13	15.4	0.28
2	4.73	0.094	14	18.8	0.17
3	5.13	0.16	15	18.8	0.22
4	5.20	0.27	16	18.8	0.28
5	5.65	0.34	17	19.2	0.32
6	5.66	0.46	18	20.0	0.33
7	7.33	0.40	19	20.5	0.37
8	9.31	0.32	20	20.5	0.43
9	9.32	0.41	21	21.1	0.45
10	14.1	0.17	22	31.4	0.09
11	14.1	0.23	23	31.6	0.11
12	15.3	0.23	24	50.7	0.001

that our dependent variable was the number of 411 calls, in thousands. We have estimated that the charge was associated with a permanent drop in the number of 411 calls by 39,500, $p < 00.001$.

7

ARIMA with Continuous Explanatory Variables

In Chapter 6, we introduced simple interventions into our ARIMA models. We saw that as long as we could assume that the ARIMA (stochastic) portion of the series we were modeling was unaffected by the intervention, we could use the ARIMA model as a baseline against which to evaluate the intervention's impact. In practical terms, this means we are assuming that the intervention affects neither the omitted independent variables that we relegated by construction to the ARIMA piece nor the structure of any random autocorrelation. This affords us the remarkable capability of providing some control for the omitted variables even when we do not know what they are and/or cannot measure them. It is a partial explanation behind the finding cited in Chapter 1 that shook the time series community: atheoretical ARIMA models were providing more accurate forecasts than multiple equation models that carefully took theory into consideration.

Our atheoretical ARIMA models are at best, though, an approximation for the variation of the omitted variables, and the approximation is based on the assumption that the omitted variables can be represented by the same model over time. Explicitly modeling the variables we are able to model is always preferred. This chapter introduces the methodology for doing so when our explanatory variable is continuous, and the extra steps that must be taken to model the dynamic relationship and minimize any spurious effects. As in Chapter 6, transfer functions will be used to model the response of our dependent variable to changes in the independent variable. Because our independent variable takes on values other than 0 and 1, though, it is no longer possible to make good guesses about the best transfer function simply by examining a plot of the series. A new tool is required: the cross-correlation function (CCF).

A second complication introduced by the inclusion of continuous explanatory variables is the likelihood of autocorrelation in the continuous explanatory

Applied Time Series Analysis for the Social Sciences: Specification, Estimation, and Inference, First Edition. Regina M. Baker.

Companion website: www.wiley.com/go/baker

variable. In the same way that two nonstationary series will be spuriously correlated, two series that share the same autocorrelation structure will be spuriously correlated; remember that autocorrelation is frequently driven by omitted variables. An important step in using the CCF to identify candidate transfer functions, then, is identifying the autocorrelation in the explanatory variable and controlling for it. In the ARIMA context, this is done through a process called "prewhitening."

The chapter proceeds as follows. Section 7.1 introduces the CCF and explains how it is used to make preliminary assessments of the response function. Section 7.2 covers prewhitening. As a practical matter, the process is often unwieldy and requires assumptions that are frequently not met in practice. This chapter, which concludes the coverage of classical time series methods, will end with a replication of early work examining the Lydia Pinkham sales and advertising series using a univariate ARIMA model, highlighting shortcomings that motivate recent developments such as the vector autoregression (VAR) approach covered later in the book.

7.1 The Cross-Correlation Function (CCF)

Relationships between two stationary time series have characteristic CCFs in the same way that within-series correlations have characteristic ACFs and PACFs. Analogous to the way that we use sample ACFs and PACFs from a series of interest to make preliminary guesses about the series' correlations with different lags of itself, a sample CCF is used to make preliminary guesses about the appropriate transfer function expressing the relationship between two series. In this section, we will define the CCF and then show why we expect it to be helpful in identifying the appropriate transfer function. Section 7.1.3 provides illustrations of transfer functions indicated by particular CCFs, and Section 7.3 provides an example using the Lydia Pinkham series.

7.1.1 Definition and Properties

If X_t and Y_t are stationary series, the CCF is defined as the set of $\rho_{XY}(k)$ such that

$$\rho_{XY}(k) = \frac{E[(X_t - \mu_X)(Y_{t+k} - \mu_Y)]}{\sigma_X \sigma_Y},$$

where

- k is an integer – that is, $k \in \{\ldots -2, -1, 0, 1, 2, \ldots\}$,
- μ_X is the mean of the series assumed to be the explanatory variable, and
- μ_Y is the mean of the series assumed to be the dependent variable.

Alternatively, we can write

$$\rho_{XY}(k) = \frac{\gamma_k(X_t, Y_{t+k})}{\sigma_X \sigma_Y},$$

where $\gamma_k(X_t, Y_{t+k})$ denotes the covariance between X_t and Y_{t+k}.

Notice the similarity between this formula and the formula for the ACF in Chapter 4. In the same way that the ACF is built from the correlations between a variable and itself in different periods, the CCF is built from the correlations between a variable X_t and a second variable Y_t in different periods. While the ACF represents within-series correlations, the CCF represents between-series correlations.

We typically write the formula in terms of X_t and future periods of Y_t to represent our expectation that X_t is the independent variable and Y_t is the dependent variable; changes in X_t are expected to precede associated changes in Y_t.[1]

Then, for example,

$$\begin{aligned}
\rho_{XY}(-2) &= corr(X_t, Y_{t-2}) \\
\rho_{XY}(-1) &= corr(X_t, Y_{t-1}) \\
\rho_{XY}(0) &= corr(X_t, Y_t) \\
\rho_{XY}(+1) &= corr(X_t, Y_{t+1}) \\
\rho_{XY}(+2) &= corr(X_t, Y_{t+2}).
\end{aligned}$$

Three important notes are in order. First, notice that the CCF is defined this way only for two stationary series. If either is nonstationary, it must be transformed and the CCF between the two stationary series is examined instead.

Second, the CCF will be useful for identifying the appropriate transfer function only if X_t, the series presumed to be the explanatory variable, is a white noise process. In practice, that is rarely the case, and so it will be necessary to "prewhiten" the series before calculating the CCF. This process is explained later in the chapter.

Finally, an important difference between the CCF and ACF is that the CCF is not symmetric. While $ACF(k) \equiv ACF(-k)$, by construction, no such symmetry is built into CCFs. In fact, asymmetry is an important feature that provides insight into the direction of causality. If, for example, X_t has an impact on Y two periods later, that is, on Y_{t+2}, we expect to see a nonzero $CCF(2)$. A nonzero $CCF(-2)$, on the other hand, would indicate a correlation between X_t and Y *two periods earlier*, that is, Y_{t-2}, which would instead be consistent with Y_t being the predictor and X_t being the dependent variable.

---

1 At first glance, it may appear that we are failing to consider anticipatory behavior – for example, making a purchase before an expected rise in price. This is not an exception to our claim that effects occur at or before the time of their causes, though, because the *anticipation* of the price change occurred at or before the time of the purchase. Indeed, the price change might not occur at all.

7.1.2 Why You Might Expect the CCF to be Helpful in Identifying the Appropriate Transfer Function

To see why this is the case, and why we might expect the CCF to be helpful in identifying the appropriate transfer function, let us derive a transfer function from a particular DGP.

Suppose $Y_t = \omega_0 X_{t-2} + N_t$, where

- X_t is free of autocorrelation,
- N_t is the ARIMA (stochastic) portion of the model, and
- N_t is independent from X_{t-2}, and with no loss of generality, let $\mu_X = \mu_Y = 0$.

Then we can derive the elements of the CCF, $\rho_{XY}(k)$, as follows.

$$\begin{aligned}\rho_{XY}(0) &= \frac{E[(X_t - \mu_X)(Y_t - \mu_Y)]}{\sigma_X \sigma_Y} \\ &= \frac{E(X_t Y_t)}{\sigma_X \sigma_Y} \\ &= \frac{E(X_t(\omega_0 X_{t-2} + N_t))}{\sigma_X \sigma_Y} \\ &= \frac{\omega_0 E(X_t X_{t-2}) + E(X_t N_t)}{\sigma_X \sigma_Y}.\end{aligned}$$

Since X_t is free of autocorrelation, $E(X_t X_{t-2}) = 0$. $E(X_t N_t) = 0$ based on one of the underlying assumptions behind ARIMA intervention models; the explanatory and noise portions of the models are independent. Thus, $CCF(0) = \rho_{XY}(0) = 0$.

The same will be true for the CCF at +1:

$$\begin{aligned}\rho_{XY}(1) &= \frac{E[(X_t - \mu_X)(Y_{t+1} - \mu_Y)]}{\sigma_X \sigma_Y} \\ &= \frac{E(X_t Y_{t+1})}{\sigma_X \sigma_Y} \\ &= \frac{E(X_t(\omega_0 X_{t-1} + N_{t+1}))}{\sigma_X \sigma_Y} \\ &= \frac{\omega_0 E(X_t X_{t-1}) + E(X_t N_{t+1})}{\sigma_X \sigma_Y} \\ &= 0.\end{aligned}$$

If you have been following the derivations carefully, you may have predicted that the pattern changes for the CCF at +2:

$$\begin{aligned}\rho_{XY}(2) &= \frac{E[(X_t - \mu_X)(Y_{t+2} - \mu_Y)]}{\sigma_X \sigma_Y} \\ &= \frac{E(X_t Y_{t+2})}{\sigma_X \sigma_Y}\end{aligned}$$

$$
\begin{aligned}
&= \frac{E[X_t(\omega_0 X_t + N_{t+2})]}{\sigma_X \sigma_Y} \\
&= \frac{\omega_0 E(X_t^2) + E(X_t N_{t+2})}{\sigma_X \sigma_Y} \\
&= \frac{\omega_0 \sigma_X^2}{\sigma_X \sigma_Y} \\
&= \omega_0 \frac{\sigma_X}{\sigma_Y}.
\end{aligned}
$$

So far, we have shown that for the series $Y_t = \omega_0 X_{t-2} + N_t$,

- $CCF(0) = 0$
- $CCF(1) = 0$
- $CCF(2) = \omega_0 \frac{\sigma_X}{\sigma_Y}$

As long as X_t is a white-noise process, it should be clear that since $E(X_t Y_{t+k}) = 0$ for all $k \neq 2$, $CCF(k)$ for $k \neq 2$ will also be zero.

It may be surprising, at first, that we do not expect a spike at lag $k = -2$, given our experience with ACFs and PACFs. To understand the asymmetry, consider the CCF at $k = -2$:

$$
\begin{aligned}
\rho_{XY}(-2) &= \frac{E[(X_t - \mu_X)(Y_{t-2} - \mu_Y)]}{\sigma_X \sigma_Y} \\
&= \frac{E[X_t Y_{t-2}]}{\sigma_X \sigma_Y} \\
&= \frac{E[X_t(\omega_0 X_{t-4} + N_{t-2})]}{\sigma_X \sigma_Y} \\
&= \frac{\omega_0 E(X_t X_{t-4}) + E(X_t N_{t-2})}{\sigma_X \sigma_Y} \\
&= 0.
\end{aligned}
$$

Instead of an expression involving $E[X_t X_t]$, we have $E[X_t X_{t-4}]$, which is 0 because of the assumption that X_t is free of autocorrelation.

In summary, for $Y_t = \omega_0 X_{t-2} + N_t$,

- $CCF(k) = 0$ for $k < 2$,
- $CCF(k) = \omega_0 \frac{\sigma_X}{\sigma_Y}$ for $k = 2$, and
- $CCF(k) = 0$ for $k > 2$.

Visually, the CCF will have a single spike at $k = +2$. The sample CCF plot for simulated data where

- $Y_t = 2X_{t-2} + a_t$,
- X_t is free of autocorrelation by construction, and
- a_t is white noise is presented in Figure 7.1.

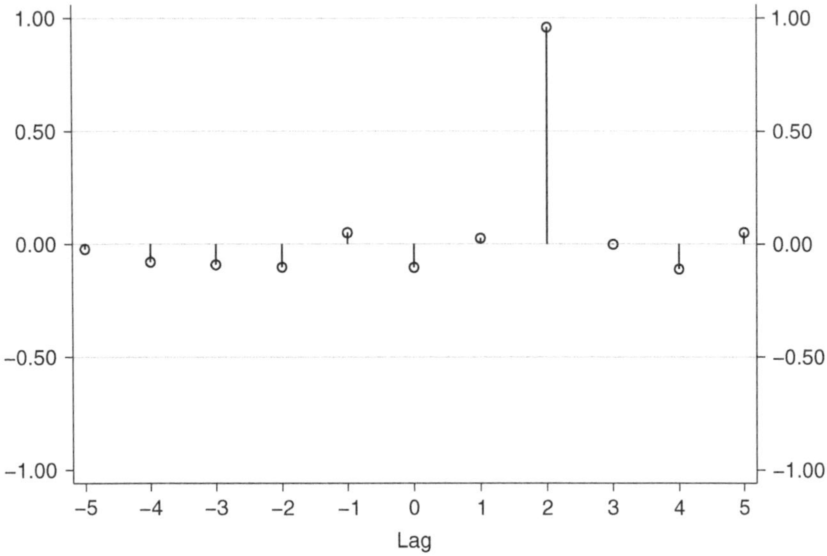

Figure 7.1 CCF for simulated $Y_t = 2X_{t-2} + a_t$, X_t white noise.

7.1.3 Using the CCF to Identify the Appropriate Transfer Function

The general form of transfer functions in the case of continuous variables is exactly the same as the general form of transfer functions for simple (0/1) interventions.

$$\upsilon(I_t) = \frac{\omega(L^s)L^l}{\delta(L^r)}, \tag{7.1}$$

where

$$\omega(L) = \omega_0 - \omega_1 L - \omega_2 L^2 - \cdots - \omega_s L^s, \tag{7.2}$$

and

$$\delta(L) = 1 - \delta_1 L - \delta_2 L^2 - \cdots - \delta_r L^r. \tag{7.3}$$

That is,

- r indicates the order of the numerator term,
- s indicates the order of the denominator term, and
- l indicates the timing of the onset of the response.

In the case of nonbinary independent variables, though, the effect of a one-unit change on the dependent series over time is not apparent from a plot of the series itself; instead, one must examine the CCF.

CCFs associated with several common transfer functions are presented in Figure 7.2.

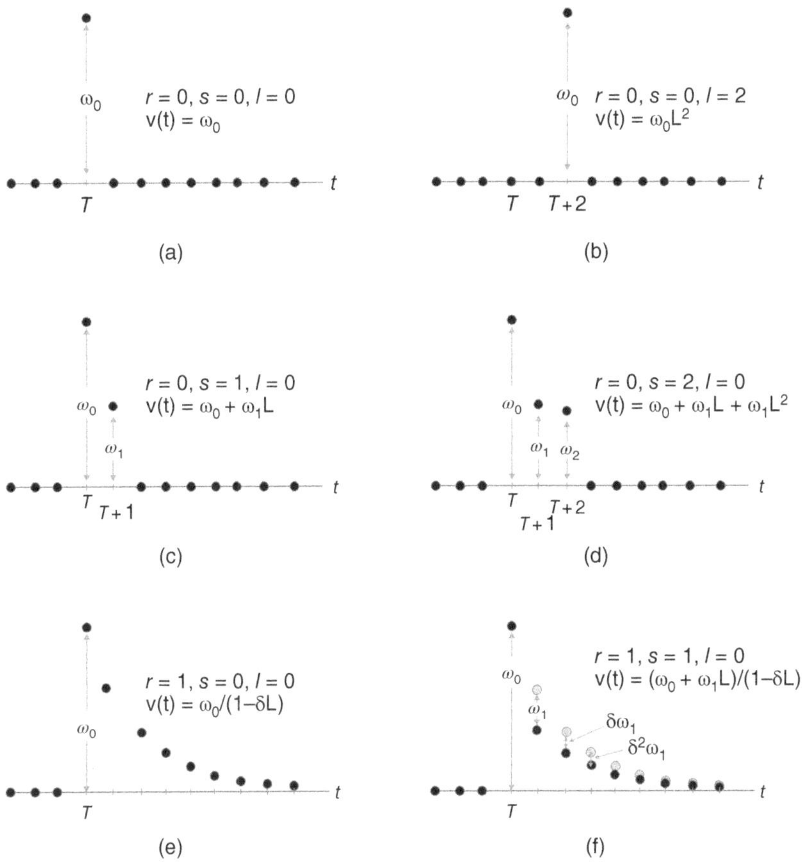

Figure 7.2 CCFs associated with commonly-used transfer functions.

As in the case of simple interventions, adding numerator elements (ω's) allows us to shift the level of the series in subsequent periods by fixed amounts. Denominator elements (δ's) permit us to model gradual changes.

Remember, though, that for the CCF to display the patterns shown, the input variable X_t must be a white noise series. If it is not, then the CCF will be muddied by indirect effects. To illustrate, suppose that in our example above X_t is correlated with X_{t-1}. Then the $CCF(1)$ and $CCF(3)$ above will not be zero anymore.

It is unusual, however, for explanatory variables of interest to be white noise series – the equivalent of being randomly distributed over time. Typically, then, the series must be transformed, and the transformed variables are used to compute the CCF, a process called "prewhitening." The prewhitened series are used only for the CCF computation; the estimation process is not affected.

7.2 Prewhitening

Intuitively, we are going to "filter" X_t's serial dependencies from *both* X_t and Y_t and then look at the CCF between the two "prewhitened" series.

Recall that any series can be modeled as an ARIMA process, so in general, X_t can be rewritten as

$$X_t = \frac{1-\theta_1 L-\theta_2 L^2-\cdots-\theta_q L^q}{-\phi_1 L-\phi_2 L^2-\cdots-\phi_p L^p} a_t,$$

where p indicates the AR order of the series and q represents the MA order. To get the white-noise portion of X_t, we invert the process:

$$\frac{1-\phi_1 L-\phi_2 L^2-\cdots-\phi_p L^p}{1-\phi_1 L-\theta_2 L^2-\cdots-\theta_q L^q} X_t = a_t,$$

and use the estimate of this a_t in our CCF computation.

To preserve the relationship between X_t and Y_t to get the correct CCF, we must make exactly the same transformation to the Y_t series – that is, we must "pre-whiten" Y_t also:

$$\text{Prewhitened } Y_t = \frac{1-\phi_1 L-\phi_2 L^2-\cdots-\phi_p L^p}{1-\theta_1 L-\theta_2 L^2-\cdots-\theta_q L^q} Y_t$$

where p and q are the AR and MA orders of the X_t process. The CCF of the two pre-whitened series will be untainted by serial correlation in X_t and will facilitate identification of the transfer function.

7.2.1 A Theoretical Illustration of Prewhitening

Let us return to the example presented earlier in the chapter, $Y_t = 2X_t + a_t$, but drop the assumption that X_t is a white noise series. Instead, let X_t be an AR(1) process with $\phi = 0.7$ – that is,

$$X_t = 0.7X_{t-1} + a_t$$
$$(1-0.7L)X_t = a_t.$$

Then

- $(1-0.7L)X_t$ is the "prewhitened" X_t, and
- $(1-0.7L)Y_t$ is the "prewhitened" Y_t.

In other words, we will compute two new series, $X_t - 0.7X_{t-1}$ and $Y_t - 0.7Y_{t-1}$, and use these to construct the CCF. For purposes of illustration, CCFs constructed

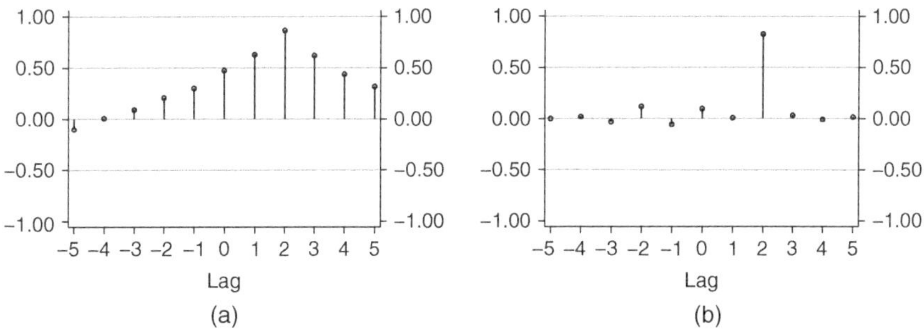

Figure 7.3 CCFs for $Y_t = 2X_{t-2} + a_t, X_t$ AR(1), with and without prewhitening. (a) CCF without prewhitening and (b) CCF with both series prewhitened.

both without and with appropriate prewhitening are presented in Figure 7.3. Notice that with the appropriate prewhitening, as in Figure 7.3a, the CCF appropriately reflects the relationship between X_t and Y_t. Without prewhitening, the CCF is muddied by indirect effects driven by the autocorrelation in X_t.

7.3 Example: Lydia Pinkham Advertising

Lydia Pinkham was nineteenth-century business pioneer whose Vegetable Compound was a precursor to modern herbal remedies and was also novel for addressing particular medical needs of women. Data for her company's sales and advertising from 1907 to 1960 were released in response to a court case. The data are regarded as particularly useful for examining the relationship between sales and advertising because her company sold only one product, and the product had no competitors. Researchers have taken a variety of different approaches to modeling the relationship, including ARIMA, a Koyck model, an error correction model, and a VAR model. For our purposes, then, it is especially useful as a means of comparing different time series methods. Here we will sketch the ARIMA approach pursued in Helmer and Johansson (1977), except that we will not exclude the last 15 years of data.[2]

The series are presented in Figure 7.4. Neither series appears to exhibit a long-term trend in mean. The long "hills and valleys," however, are often seen in series with unit roots; indeed, the series underscore the importance of testing for

2 Using the computational methods available in 1977, Helmer and Johansson dropped the last 15 data points from their analysis so that they could be used later to compare ARIMA forecasts with forecasts from other models.

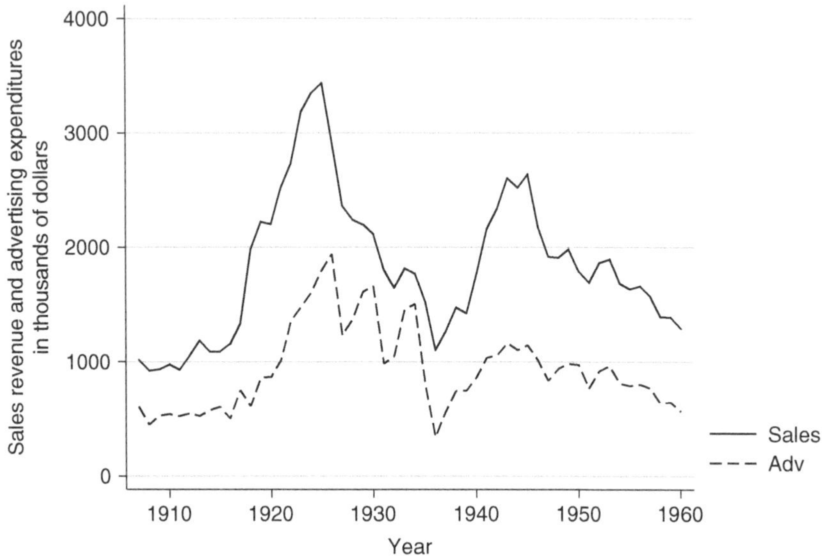

Figure 7.4 Sales revenue and advertising expenditures for Lydia Pinkham's Vegetable Compound, 1907–1960.

non-stationarity even when there is no apparent long-term trend. Dickey–Fuller tests indicate unit roots in both, and for both, first differencing is sufficient for achieving stationarity.

Like Helmer and Johansson, we begin with the hypothesis that advertising drives sales, and so our first step is to determine whether the differenced advertising series is free of autocorrelation; if not, we will need to prewhiten both series before calculating the CCF.

The sample ACF for the differenced advertising series is presented in Figure 7.5. The alternating spikes at lags 2 and 4 are common characteristics of AR(2) models and indeed, Helmer and Johansson settled on an AR(2) model for the differenced advertising series. Our estimates of the autoregressive coefficients for an AR(2) model using all data are very similar to the estimates they obtained for their shortened series, $\hat{\varphi}_1 = 0.048$ and $\hat{\varphi}_2 = -0.387$, with estimated standards errors 0.130 and 0.129, respectively.[3]

The residuals from our model are presented in Figure 7.6. The statistically significant spike at lag 4, as well as smaller spikes at 2 and 6, is troubling. A test of the

3 Helmer and Johansson estimated $\hat{\varphi}_1 = 0.074$ and $\hat{\varphi}_2 = -0.407$ dropping the last 15 observations of the series.

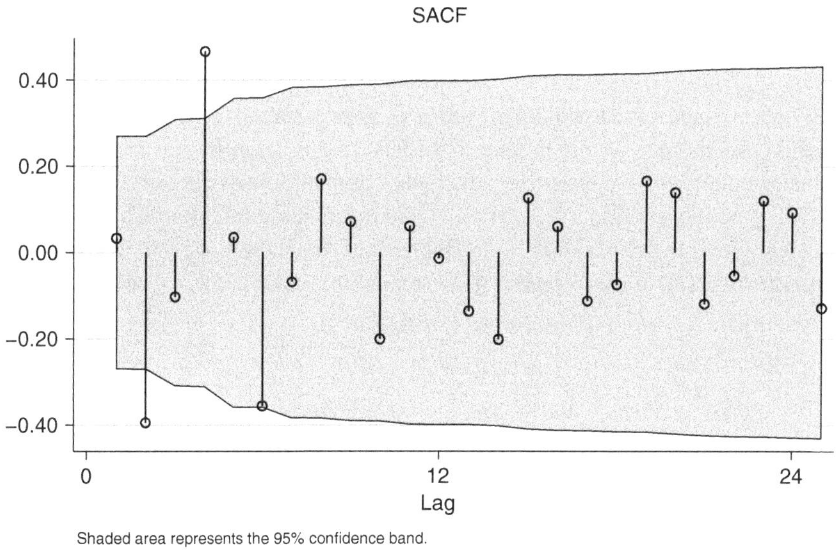

Figure 7.5 ACF for differenced advertising series.

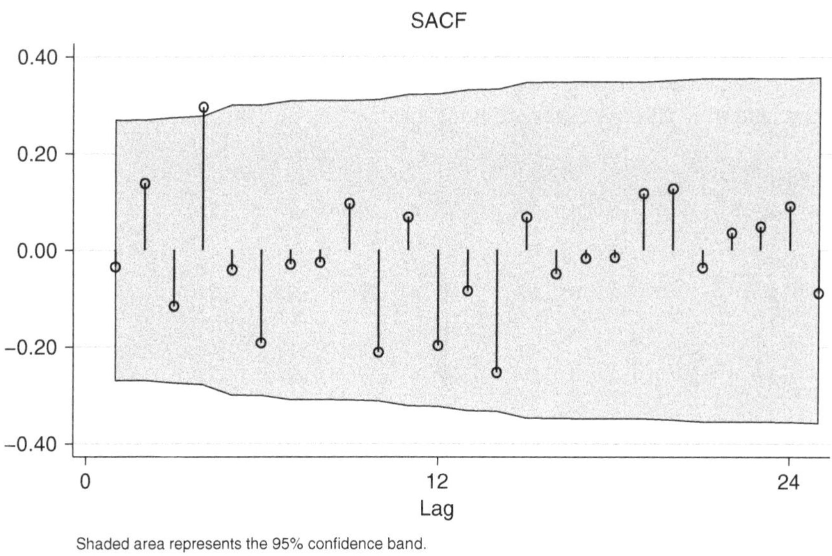

Figure 7.6 ACF for residuals from (2,1,0) model for Lydia Pinkham's advertising series.

null hypothesis that autocorrelations at all of the first six lags are 0 using a Q-test yields $Q(4) = 9.57$, $p = 0.048$; that is, we cannot reject the null hypothesis of no remaining autocorrelation at the 0.05 level of significance.

If our end-goal were to model the advertising series, this would be a minor issue. In this case, though, we are modeling the advertising series to find an appropriate transformation to prewhiten the advertising and sales series before computing a CCF. It is possible to find a (2,1,2) model that meets standard criteria, but this is too unwieldy for a transformation. Like Helmer and Johansson, we will instead work with transformation based on the AR(2) model instead. That is, we compute

$$\Delta Adv_t^* = \Delta Adv_t - 0.048\Delta Adv_{t-1} + 0.387\Delta Adv_{t-2},$$

and

$$\Delta Sales_t^* = \Delta Sales_t - 0.048\Delta Sales_{t-1} + 0.387\Delta Sales_{t-2},$$

and compute the CCF using these transformed series.

The resulting CCF is presented in Figure 7.7. Here we receive some troubling news. While the CCF for lags $k \geq 0$ look very similar to those Helmer and Johansson presented (1977:233), we also see a large spike at lag $k = -1$, a portion of the CCF that Helmer and Johansson did not include. We cannot be sure, given the limitations of our prewhitening, but the cross-correlation at $k = -1$ suggests that we may have reversed or reciprocal causality: we need to consider the possibility that sales drives advertising. Indeed, Helmer and Johansson acknowledge the possibility that advertising budgets are driven in part by sales revenues in previous years, and they propose other methods for investigating this possibility.

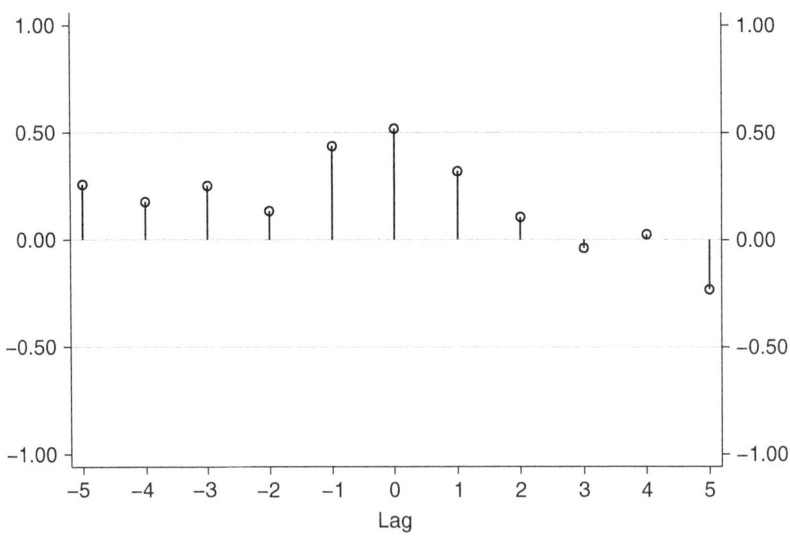

Figure 7.7 CCF for prewhitened differenced sales and advertising series.

7.4 Conclusion

In this chapter, we have seen that it is possible to develop ARIMA models with continuous explanatory variables, but the process is sometimes more cumbersome in practice than it might appear in theory. CCFs provide useful information for making preliminary guesses about transfer functions if the autocorrelation structure in the explanatory variable can be represented by a simple ARIMA process, but otherwise the transformations required for prewhitening can become unwieldy, and without starting points, finding good models can involve a great deal of trial and error.

The CCF also provides a powerful reminder of the weighty assumption we make when we estimate single-equation models – that there is a single direction of causality, and that we know what it is. VAR models, which permit the estimation of reciprocal causality, were introduced seven years after the publication of Helmer and Johansson's ARIMA model of Lydia Pinkham's sales and advertising data. The concern that the standard econometric approaches of the time were too "theory-laden" – too reliant on untested assumptions – underpins both the general-to-specific modeling approaches of the LSE school and the VAR approach introduced by Christopher Sims. Both combine insights of classical time series modeling with elements of the traditional regression approach, so we will review the latter before proceeding.

Reference

Helmer R.M. and Johansson J.K. (1977). An Exposition of the Box-Jenkins Transfer Function Analysis With an Application to the Advertising-Sales Relationship. *Journal of Marketing Research* XIV(May):227–239.

8

OLS and the Gauss–Markov Assumptions

Let us re-examine what we learn about autocorrelation in introductory regression courses now that we know a little more about time series properties. This chapter introduces critiques both of standard methods for detecting autocorrelation, and of reflexively "correcting" autocorrelation with generalized least squares (GLS) approaches. While the advice in this chapter is not limited to the contents of Grayham Mizon's 1995 paper, its gist is well-summarized by the title: "A simple message for autocorrelation-correctors: Don't."

The first section explaining autocorrelation and its consequences should be review for most readers. The second section introduces both the Durbin–Watson and Breusch–Godfrey methods of detecting autocorrelation, and evaluates the advantages and disadvantages of these statistics compared with tools like the autocorrelation function and Q statistics presented in Chapter 4.

The standard "fix" – approaches based on GLS – is presented in the third section. While a typical matrix algebra derivation is provided, a simpler presentation using lag operator algebra is also introduced and used to derive extensions of the model. The chapter concludes with two caveats regarding these approaches. The "Common Factors" critique is a reminder that the assumptions underpinning GLS methods are restrictive, and that the standard GLS model correcting for first-order autocorrelation is just one special case of the autoregressive distributed lag (ADL) models presented in more detail in Chapters 9 and 10. Finally, alternative models testing hypotheses concerning the factors driving Iraqi civilian deaths in the recent Iraq War are used to illustrate an important theme in the book – that autocorrelation is sometimes better addressed by thinking more carefully about model specification.

Applied Time Series Analysis for the Social Sciences: Specification, Estimation, and Inference, First Edition. Regina M. Baker.

Companion website: www.wiley.com/go/baker

8.1 Autocorrelation and its Consequences: A Review

Beginning with the simple model

$$Y_t = \beta_0 + \beta_1 X_t + u_t, \tag{8.1}$$

where $u_t \sim N(0,\sigma^2)$,
we are taught the following regarding autocorrelation:

- By definition, autocorrelation occurs when error at time t is correlated with the error term from a previous time period – for example, when omitted causes of Y at time t also affect Y in the previous (or next) period.
- Autocorrelation is a violation of the Gauss–Markov assumption that $E(u_t u_s) = 0,\ t \neq s$.
- Because this assumption enters only into the derivation of the standard error of the estimate, σ_{β_1}, and not into the proof that the OLS regression estimate $\hat{\beta}$ is unbiased, coefficient estimates from OLS estimations with autocorrelation will be unbiased but standard errors are suspect.

To better understand the logic behind both GLS and Newey–West estimators, it is worth reviewing the problem in matrix form. Beginning with

$$\mathbf{Y} = \mathbf{X}\beta + \mathbf{U} \tag{8.2}$$

where

- $\mathbf{Y}$ is a $T \times 1$ column vector[1] representing the dependent variable,
- $\mathbf{X}$ is the $T \times (k+1)$ independent variable matrix, where k is the number of independent variables, and a constant term is included,
- β is a $(k+1) \times T$ vector of coefficients to be estimated, and
- $\mathbf{U}$ is the $T \times 1$ vector of error terms.

In the absence of heteroskedasticity and autocorrelation,

$$\mathrm{Var}\left[\hat{\beta} \mid \mathbf{X}\right] = E(\mathbf{X}'\mathrm{X})^{-1}\mathrm{X}'UU'\mathrm{X}(\mathbf{X}'\mathrm{X})^{-1} \mid \mathbf{X}] \tag{8.3}$$

$$= E(\mathbf{X}'\mathrm{X})^{-1}\mathbf{X}'\sigma^2\mathbf{I}\mathrm{X}(\mathbf{X}'\mathrm{X})^{-1} \mid \mathbf{X}], \tag{8.4}$$

where $\mathbf{I}$ is the appropriately dimensioned identity matrix.

---

1 If you are relatively new to matrix algebra, drawing diagrams representing the shapes of the matrices is often a useful exercise to help you associate the correct visual image with the matrix notation.

The step from Equation (8.3) to Equation (8.4) is valid, however, only if the expression in the center, $\mathbf{UU}'$, is $\sigma^2\mathbf{I}$ in expectation, and this is true only if the error terms are homoskedastic and not autocorrelated. In the case of heteroskedasticity but no autocorrelation,

$$E\left(\mathbf{UU}'\right)=\begin{bmatrix} E\left(u_1^2\right) & E\left(u_1u_2\right) & \cdots & E\left(u_1u_T\right) \\ E\left(u_2u_1\right) & E\left(u_2^2\right) & \cdots & E\left(u_2u_T\right) \\ \vdots & \vdots & \ddots & \vdots \\ E\left(u_Tu_1\right) & E\left(u_Tu_2\right) & \cdots & E\left(u_T^2\right) \end{bmatrix} = \begin{bmatrix} \sigma_1^2 & 0 & \cdots & 0 \\ 0 & \sigma_2^2 & \cdots & 0 \\ \vdots & \vdots & \ddots & \vdots \\ 0 & 0 & \cdots & \sigma_T^2 \end{bmatrix} \tag{8.5}$$

where $\sigma_s^2 \neq \sigma_t^2$ for some $s \neq t$.

In the case of autocorrelation, the off-diagonal elements of $E[\mathbf{UU}']$ are no longer 0.

Since autocorrelation is almost always positive, autocorrelation typically leads to underestimation of standard errors, predisposing practitioners to infer that coefficients are statistically significant when they should not. GLS (or weighted least squares) can be used to correct the standard errors. Since it is only a re-weighting scheme, the coefficient estimates from GLS are still unbiased, on average.[2]

An important theme in Chapter 1 is worth reiterating here. Recall the entire set of Gauss–Markov assumptions:

- $E(\mathbf{X}'\mathbf{U}) = \mathbf{0}$ – i.e., we have not omitted any confounding variables.
- $\mathbf{X}$ is fixed in repeated samples.
- $E(u_t) = 0$.
- $\text{Var}(u_t) = \sigma^2$ – i.e., the error term is homoskedastic; its variance is constant across observations).
- $E(u_s u_t) = 0, s \neq t$ (no autocorrelation).

2 It is worth remembering what "on average" means here: that the parameter estimates from all possible realizations of our process will center around the process's true parameters. In practice, we typically have only one realization. In situations where there are individual observations that are influential (e.g., because they are outliers), different weighting schemes may lead to very different point estimates.

The common wisdom is based on the assumption that *only* the "no autocorrelation" assumption is violated. Autocorrelation, though, is frequently caused by omitted variables. To believe that biased coefficients are not a concern when autocorrelation is detected requires the rather heroic assumption either that the autocorrelation is not driven by omitted variables or other misspecification, or that any omitted variable or misspecification driving the autocorrelation is not associated with both the dependent variable and the included independent variables. If one or the other assumption is not met, $E(\mathbf{X}'\mathbf{U}) = \mathbf{0}$ is violated also, introducing omitted variables bias. Generalized least squares approaches do not fix this.

> The presence of autocorrelation should raise concerns not only about incorrectly estimated standard errors, but also about incorrectly estimated coefficients. Addressing problems with the model specification, when possible, is the best way to fix autocorrelation.

8.2 Detecting Autocorrelation

Suppose you have estimated the model described in Equation (8.1). First cuts at detecting autocorrelation include plotting the estimated residual terms, $\hat{u}_t$, against $\hat{u}_{t-1}$ and/or regressing $\hat{u}_t$ against $\hat{u}_{t-1}$. In the absence of autocorrelation, no pattern will be observed in the plot, and the estimated regression coefficient will be 0 or very close to it.

8.2.1 The Durbin–Watson Test

The most commonly presented diagnostic tool is the Durbin–Watson statistic, d, estimated from the residuals as

$$\hat{d} = \frac{\sum_{t-2}^{T} (\hat{u}_t - \hat{u}_{t-1})^2}{\sum_{t=1}^{T} \hat{u}_t^2} \tag{8.6}$$

$$= \frac{\sum_{t=2}^{T} (\hat{u}_t^2 = 2\hat{u}_t \hat{u}_{t-1} + \hat{u}_{t-1}^2)}{\sum_{t=1}^{T} \hat{u}_t^2}. \tag{8.7}$$

Notice that as the number of time points T becomes very large,

$$\sum_{t=2}^{T} \hat{u}_t^2 \approx \sum_{t=2}^{T} \hat{u}_{t-1}^2,$$

so Equation (8.7) is approximately equal to

$$\frac{2\sum_{t=2}^{T}\hat{u}_t^2 - 2\hat{u}_t\hat{u}_{t-1}}{\sum_{t=1}^{T}\hat{u}^2}$$

$$= 2 - 2\frac{\sum_{t=2}^{T}\hat{u}_t\hat{u}_{t-1}}{\sum_{t=1}^{T}\hat{u}_t^2}$$

From the derivations of autocorrelation functions in Chapter 3, though, you will recognize that this expression is equivalent to $2-2\hat{\rho}$, where ρ is the correlation between u_t and u_{t-1}. In other words,

$$\hat{d} \approx 2 - 2\hat{\rho}.$$

This leads to boundaries on the value of the Durbin–Watson statistic:

- If $\hat{\rho}=1, \hat{d} \approx 2-2(1)=0$. In the case of perfect positive first-order autocorrelation, the DW statistic is 0.
- If $\hat{\rho}=0, \hat{d} \approx 2-2(0)=2$. In the case of no first-order autocorrelation, the DW statistic is 2.
- If $\hat{\rho}=-1, \hat{d} \approx 2-2(-1)=4$. In the case of perfect negative first-order autocorrelation, the DW statistic is 4.

That is, the DW statistic ranges from 0 to 4. In the absence of first-order autocorrelation, the DW statistic is expected to be 2.

If the DW statistic provides evidence of autocorrelation, ρ can be estimated directly from the DW statistic:

$$\hat{\rho} \approx 1 - \frac{\hat{d}}{2}. \tag{8.8}$$

8.2.1.1 Limitations of the Durbin–Watson Statistic

One limitation of the widely-used DW statistic should be obvious already: it is a test only for first-order autocorrelation.[3] It is also appropriate only for models that include a constant term, and it is not appropriate for models that include a lagged dependent variable[4] – e.g.,

$$Y_t = \beta_0 + \beta_1 Y_{t-1} + \cdots + u_t.$$

---

3 There are, however, generalized Durbin–Watson statistics that can be computed in some statistical packages.

4 For large samples, Durbin's h statistic may be appropriate in this case.

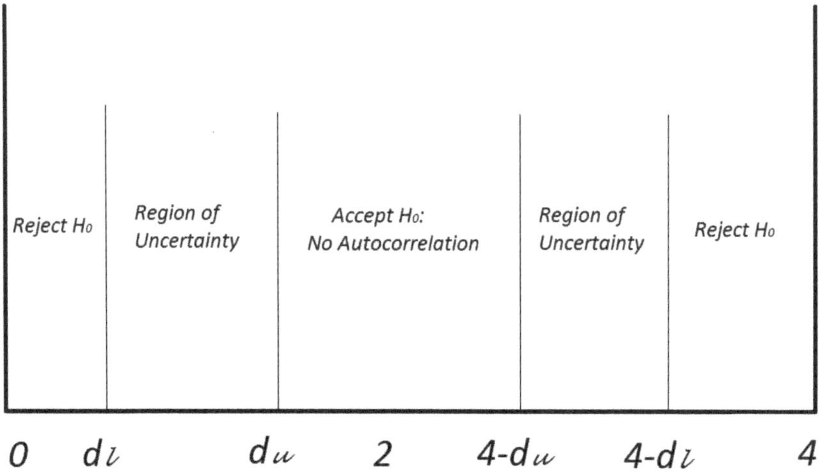

Figure 8.1 Regions of uncertainty for the Durbin–Watson statistic.

Notice, though, that the autocorrelation function (ACF) and associated Q-statistics are much more informative. While the DW statistic detects only first-order autocorrelation, the ACF provides information about autocorrelation for several lags.

More debilitating than any of these limitations, though, is the nature of the DW statistics' distribution. The estimated DW and its associated degrees of freedom will not always permit hypothesis testing, but instead will provide the values d_l and d_u associated with the graphic in Figure 8.1.

If the estimated DW statistic falls between d_l and d_u, or between $4-d_u$ and $4-d_l$, the test is inconclusive; one can neither reject any null hypothesis nor fail to reject a null hypothesis. The distribution of the parameter is simply unknown in these instances. These regions of uncertainty become larger as the degrees of freedom for the test become smaller.

8.2.2 Breusch–Godfrey Tests

Breusch–Godfrey tests are a form of Lagrange-Multiplier test, and much more informative than the DW test. While they are now implemented in many

5 The figure multiplied by R^2 is the number of cases used in the regression generating the R^2 used in the test statistic. An alternate approach is to replace missing values of the lagged $\hat{u}_t$ with 0, in which case the u_t regression can be run using T cases, and the test statistic becomes TR^2.

statistical packages, they can also be computed easily using any package that permits regression computations using the following algorithm.

- Regress the dependent variable on the independent variables and save the estimated residuals, the $\hat{u}_t$'s.
- Regress u_t on all the independent variables and lags of the estimated residuals, $\hat{u}_{t-1}, \hat{u}_{t-2}, \ldots \hat{u}_{t-k}$. The number of lags to include is a judgment call; thinking about the nature of the data and/or looking at the estimated autocorrelation function of the residuals may be helpful here. For example, if the data are quarterly and/or the autocorrelation function shows important spikes at the 4th period, including lags at least as far back as $\hat{u}_{t-4}$ is indicated.
- Save the R^2 from this regression.
- If the sample size is large,

$$(T-k)R^2 \sim \chi_k^2,$$

  where
  T is the number of time points included in the original regression and k is highest order of lag included.[5]

Compared to the Durbin–Watson statistic, Breusch–Godfrey tests have these advantages:

- They can be used with lagged dependent variables (unlike the DW statistic).
- The original equation need not include a constant term.
- Breusch–Godfrey tests can be used to detect higher-order autocorrelation, although the researcher must specify the appropriate lag length to do this.

Like the Durbin–Watson statistic, though, Breusch–Godfrey tests do not provide any more information than the autocorrelation function of the estimated residuals and the associated Q-statistics.

8.2.3 Example: Civilian Deaths During the Iraq War

Let us return to the Iraq War data introduced earlier in the book. In Chapter 6, analyses involving only the number of US troops and the number of Iraqi civilian casualties provided support for the assertion that the 2007 "surge" was associated with declining sectarian violence in Iraq. Critics of this position have offered (at least) two challenges to the suggestion that the increase in troop levels were responsible for the decline. First, the reminder that there was a dramatic increase in sectarian violence following the 2006 bombing of the al-Askari mosque in Samarra suggests that this event should be taken into account: any later decline in civilian casualties might represent only a return to the levels observed prior to the al-Askari bombing in February of 2006. The models that follow include a dummy variable referred to as "Samarra" which is coded as 0 through January of 2006, and 1 afterward. Second, the "surge" – the 2007 increase in US troops deployed in Iraq – was introduced as part of revised strategy referred to as "The New Way Forward." To control for the broader change in strategy, a dummy variable representing the "New Way Forward" is included. It is coded 0 through January of 2007 and 1 after that. The results of a regression using ordinary least squares (OLS) is presented in the first column of Table 8.1. The results presented in the other columns will be discussed later in the chapter.

The OLS results do not provide support for the claim that the increase in troop deployment was associated with a decline in civilian casualties. The estimated coefficient of 5.38 is not statistically significant, and more important, it is positive rather than negative. The results do, however, provide an estimate of the average increase in the number of civilian casualties following the bombing in Samarra: monthly casualties increased by an average of 1 642, $p < 0.001$. The results also provide support for the claim that the New Way Forward was associated with a reduction in civilian casualties, even after controlling for the changes in the level of the US troop deployment. On average, there were 1963 fewer deaths each month following the change in strategy, $p < 0.001$.

Table 8.1 Regression estimates for models of Iraqi civilian deaths.

	OLS	pGLS	OLS
L(CivDeaths)			0.70***
			(0.08)
US Troops	5.38	−14.04	−2.78
	(8.76)	(10.05)	(5.91)
New Way Forward	−1963***	−877*	−708**
	(277)	(405)	(229)
Samarra	1642***	638	555
	(225)	(392)	(189)
Constant	644	3492	835
	(1230)	(1466)	(816)
n	69	69	68
SE	659	453	434
DW (4,69)	0.65		

Standard errors are reported in parentheses.
SE is the standard error of the regression.
DW is the Durbin–Watson statistic.
$^{*}p < 0.05$, $^{**}p < 0.01$, $^{***}p < 0.001$.

We know, however, that we should be wary of significance levels estimated from models with autocorrelation, and the Durbin–Watson statistic of 0.65 suggests that the error terms in the model are positively autocorrelated. From Equation (8.8), a rough estimate of the autocorrelation coefficient is

$$\hat{\rho} \approx 1 - \frac{0.65}{2}$$
$$= 0.68.$$

This point estimate is sizable, and would be worth considering even if we were unsure of its statistical significance. To evaluate the latter, though, we find the values of d_l and d_u associated with 4 included variables and 69 cases: 1.52 and 1.70, respectively. The relevant cutpoints for Figure 8.1, then, are 0, 1.52, 1.70, 2, 2.30, 2.48 and 4.0. Our estimate of 0.65 falls in the first region, between 0 and $d_l = 1.52$, leading us to reject the null hypothesis of no autocorrelation.

We would reach the same conclusion using the Breusch–Godfrey (BG) test. Testing only for first-order autocorrelation would require estimating the following regression:

$$\hat{u}_t = \beta_0 + \gamma_1 \text{USTroops} + \gamma_2 \text{NewWayForward} + \gamma_3 \text{Samarra} + \beta_1 \hat{u}_{t-1} + \epsilon_t, \quad (8.9)$$

where the β's and γ's are coefficients to be estimated, and ϵ_t is the error term. The R^2 from this estimation is 0.47 and there are 69 time points in our original analysis, so our test statistic becomes $(0.47)(69-1) = 32$. Evaluating this against the χ^2 distribution with 1 degree of freedom, we find that we can reject the null hypothesis that $\beta_1 = 0$, $p < 0.001$.

While it is reassuring to find agreement between the DW and BG tests for first-order autocorrelation, one advantage of the BG test is its ability to detect higher-order autocorrelation. To do so, we add higher order lags of $\hat{u}_t$ to the independent variables in Equation (8.9). Suppose we believed there might be quarterly effects, perhaps because troop rotations occurred on 3-month cycles. Then the new auxiliary equation is instead

$$\hat{u}_t = \beta_0 + \gamma_1 \text{USTroops} + \gamma_2 \text{NewWayForward} + \gamma_3 \text{Samarra}$$
$$+ \beta_1 \hat{u}_{t-1} + \beta_2 \hat{u}_{t-2} + \beta_3 \hat{u}_{t-3} + \beta_4 \hat{u}_{t-4} + \epsilon_t.$$

The R^2 from the estimation would be multiplied by $65(69-4)$ and compared with the χ^2 distribution with 4 degrees of freedom to test the null hypothesis that $\beta_1 = \beta_2 = \beta_3 = \beta_4 = 0$. The reader is asked to do this and compare the results to the information obtained from the ACF and associated Q-tests for $\hat{u}_t$ Exercise (8.1).

8.3 Generalized Least Squares

Once autocorrelation has been detected, the standard recommendation in introductory regression courses is to reestimate the model using a GLS-based approach. The foundations and mechanics of the approach are presented in this and following sections.

Since autocorrelation affects OLS through the variance matrix $\mathbf{UU}'$, a common recommendation is to model the autocorrelation, by assuming that the the errors follow an AR(1) process. In other words, we begin with Equation (8.1), but instead of assuming that u_t is normally, independently distributed, we model the error terms as

$$u_t = \rho u_{t-1} + \epsilon_t \tag{8.10}$$

- $-1 \leq \rho \leq 1$
- $E(\epsilon_t) = 0$
- $E(\epsilon_t \epsilon_s) = 0, \quad t \neq s$
- $E(\epsilon_t^2) = \sigma_\epsilon^2$

Two derivations of the Generalized Least Squares (GLS) estimator follow. The standard presentation is laborious; a derivation using lag operator algebra is much simpler and lays the groundwork for other insights.

8.3.1 Standard Presentation of GLS

If the autocorrelation can be modeled as Equation (8.10), then the $\mathbf{E}(\mathbf{UU}')$ matrix, Equation (8.5) becomes

$$\begin{aligned}
E(\mathbf{UU}') &= \begin{bmatrix} E(u_1^2) & E(u_1u_2) & \dots & E(u_1u_T) \\ E(u_2u_1) & E(u_2^2) & \dots & E(u_2u_T) \\ \vdots & \vdots & \ddots & \vdots \\ E(u_Tu_1) & E(u_Tu_2) & \dots & E(u_T^2) \end{bmatrix} \\
&= \begin{bmatrix} \sigma_u^2 & \rho\sigma_u^2 & \dots & \rho^{T-1}\sigma_u^2 \\ \rho\sigma_u^2 & \sigma_u^2 & \dots & \rho^{T-2}\sigma_u^2 \\ \vdots & \vdots & \ddots & \vdots \\ \rho^{T-1}\sigma_u^2 & \rho^{T-2}\sigma_u^2 & \dots & \sigma_u^2 \end{bmatrix} \\
&= \sigma_u^2 \begin{bmatrix} 1 & \rho & \dots & \rho^{T-1} \\ \rho & 1 & \dots & \rho^{T-2} \\ \vdots & \vdots & & \vdots \\ \rho^{T-1} & \rho^{T-2} & \dots & 1 \end{bmatrix}
\end{aligned}$$

Recall from Exercise 4.3 that (8.10) can be rewritten as

$$\sigma_u^2 = \frac{\sigma_\epsilon^2}{1-\rho^2}.$$

Then

$$E(\mathbf{U}\mathbf{U}') = \frac{\sigma_\epsilon^2}{1-\rho^2}\begin{bmatrix} 1 & \rho & \dots & \rho^{T-1} \\ \rho & 1 & \dots & \rho^{T-2} \\ \vdots & \vdots & \ddots & \vdots \\ \rho^{T-1} & \rho^{T-2} & \dots & 1 \end{bmatrix}$$
$$= \sigma_\epsilon^2 \Omega,$$

where

$$\Omega = \frac{1}{1-\rho^2}\begin{bmatrix} 1 & \rho & \dots & \rho^{T-1} \\ \rho & 1 & \dots & \rho^{T-2} \\ \vdots & \vdots & \ddots & \vdots \\ \rho^{T-1} & \rho^{T-2} & \dots & 1 \end{bmatrix} \quad (8.11)$$

The inverse matrix is rarely derived, but can be written as

$$\Omega^{-1} = \begin{bmatrix} 1 & -\rho & 0 & \dots & 0 & 0 & 0 \\ -\rho & 1+\rho^2 & -\rho & \dots & 0 & 0 & 0 \\ 0 & -\rho & 1+\rho^2 & \dots & 0 & 0 & 0 \\ \vdots & \vdots & \vdots & \ddots & \vdots & \vdots & \vdots \\ 0 & 0 & 0 & \dots & -\rho & 1+\rho^2 & -\rho \\ 0 & 0 & 0 & \dots & 0 & -\rho & 1 \end{bmatrix} \quad (8.12)$$

8.3.2 A Simpler Derivation Using Lag Operator Algebra

Remember we started with

$$Y_t = \beta_1 + \beta_2 X_t + u_t,$$

where

$$u_t = \rho u_{t-1} + \epsilon_t,$$

and ϵ_t was a white-noise error term.

Notice that using lag operator notation, we can rewrite the equation for the error term:

$$\begin{aligned} u_t - \rho u_{t-1} &= \epsilon_t \\ (1-\rho L)u_t &= \epsilon_t \\ u_t &= \frac{\epsilon_t}{1-\rho L}. \end{aligned}$$

Substitute this expression for u_t into the regression equation we were trying to estimate:

$$Y_t = \beta_0 + \beta_1 X_t + \frac{\epsilon_t}{1-\rho L}. \tag{8.13}$$

From your reading of Chapter 2, you will likely recognize that multiplying the equation through by $1-\rho L$, and rearranging terms, will yield an equivalent equation that can be estimated by OLS:

$$\begin{aligned} (1-\rho L)Y_t &= (1-\rho L)\beta_0 + (1-\rho L)\beta_1 X_t + \epsilon_t \\ Y_t - \rho Y_{t-1} &= (1-\rho)\beta_0 + \beta_1 X_t - \rho\beta_1 X_{t-1} + \epsilon_t \end{aligned}$$

$$Y_t = (1-\rho)\beta_0 + \rho Y_{t-1} + \beta_1 X_t - \rho\beta_1 X_{t-1} + \epsilon_t. \tag{8.14}$$

That is, the GLS model derived using matrix algebra in the previous section is mathematically equivalent to estimating Equation (8.14) using OLS. The derivation using lag operator algebra affords several advantages. First, it is much simpler. Second, it makes it easier to think about how to extend the analysis to other error structures. Examples are provided in the next section, and in exercises at the end of the chapter. Finally, it leads easily into the Common Factors Critique, presented below.

8.3.2.1 A Simple Extension

Suppose $\rho = 1$. At first blush, this might appear to be a serious problem; $\rho = 1$, after all, means that the residuals are perfectly correlated with each other. But then Equation (8.13) becomes

$$Y_t = \beta_0 + \beta_1 X_t + \frac{\epsilon_t}{1-L}.$$

Multiplying the equation through by $1-L$ and rearranging terms first yields

$$Y_t - Y_{t-1} = \beta_1 X_t - \beta_1 X_{t-1} + \epsilon_t.$$

Rewriting this as an equation in first differences gives us the equation we are advised to estimate in this case:

$$\Delta Y_t = \beta_1 \Delta X_t + \epsilon_t.$$

8.3.3 Pseudo-GLS Techniques

GLS assumes that the "true" autocorrelation coefficient, ρ, is known. This is rarely the case in practice, however. Simultaneous estimation of ρ and the β's involves solving an equation that is nonlinear in the parameters, so numerical methods are required.

8.3.3.1 The Cochrane–Orcutt Method

The Cochrane–Orcutt method is typically available as an option in standard statistical software. Behind the scenes, the simultaneous estimation of ρ and the β's involves successive iterations of roughly the following steps.

1) Estimate the initial equation of interest – for example

$$Y_t = \beta_0 + \beta_1 X_t + u_t$$

– and save the residuals.
2) Using those residuals, estimate the regression

$$\hat{u}_t = \hat{\rho}\hat{u}_{t-1} + v_t$$

to get an initial estimate of $\hat{\rho}$.
3) This $\hat{\rho}$ is used in transformations to obtain new estimates of β_0 and β_1.
4) These new estimates are used to compute new values for the error terms ϵ_t, which are, in turn, used to compute a new value for ρ.

The steps are repeated until successive estimates of ρ are close to each other.

8.3.3.2 The Prais–Winsten Correction

The Cochrane–Orcutt estimation procedure includes a step that uses lags of the dependent and independent variables, and therefore requires dropping the first case in the data set. Prais and Winsten pointed out that this can lead to problems in small samples. To avoid dropping the first case, they devised a transformation that permits it to be included. The rest of the (iterative) estimation process is exactly the same as the Cochrane–Orcutt procedure.

8.3.4 Example: Civilian Deaths During the Iraq War

The results of estimating our model of Iraqi civilian deaths using the Cochrane–Orcutt procedure with the Prais–Winsten correction are presented in the second column of Table 8.1. Notice that the standard errors (and hence, p-values) associated with all estimated coefficients are larger than those estimated with OLS. In this case, coefficient estimates are very different as well. Our estimate of the impact of the al-Askari bombing on civilian deaths has changed from an increase of 1642 casualties per month (OLS) to an increase of 638 casualties per month (pGLS), and is no longer statistically significant. Our estimate of the effect of the New Way Forward has changed from a decrease of 1963 civilian deaths per month to a decrease of 877 deaths per month. Such changes in coefficient estimates are not uncommon, unfortunately, and are cause for alarm. Remember that in expectation, the coefficient estimates of OLS and GLS (or pGLS) are identical. GLS and pGLS are re-weighting schemes, though, and estimated coefficients will differ if the weights given to particularly influential data points are changed. Instability of coefficient estimates is one sign that our model is missing something important.

8.4 Limitations of GLS Approaches

8.4.1 Dynamic Misspecification

In earlier chapters, we saw that autocorrelation is frequently caused by omitted variables, and that if these variables are associated with included variables, our coefficient estimates will be biased. This is no less true for GLS methods than it is for OLS; "fixing" the error term does not address misspecification problems.

Our study of transfer functions in Chapter 6 should also give you pause. In our model of Iraqi civilian casualties, for example, we have implicitly assumed that the al-Askari bombing led to an immediate and permanent increase in the monthly level of casualties. This seems implausible, though, and is not supported by visual inspection of the monthly casualty series. Recall that the shape of the transfer function can be modified by including a lagged dependent variable in the model specification. The final column of Table 8.1 lists the results of estimating such a regression. Examination of the residuals will show that first-order autocorrelation is no longer a problem. Just as the ω parameters in our transfer function models could not be interpreted without reference to the coefficient on the lagged dependent variable, though, the estimated coefficients in the final column of Table 8.1 must be interpreted in light of the sizable estimate of the coefficient associated with the lagged civilian deaths variable. This topic is explored in depth in Chapters 9 and 10.

8.4.2 The Common Factors Critique of GLS

Work by Hendry and Mizon, 1978 called attention to the fact that mindlessly applying the GLS model described above imposes restrictions on a more general model. The restrictions may or may not be valid, but can be tested by estimating the general model.

Suppose the "true" model is

$$Y_t = \gamma_1 + \gamma_2 X_t + \gamma_3 X_{t-1} + \gamma_4 Y_{t-1} + \varepsilon_t$$

but instead, you estimate

$$Y_t = \beta_0 + \beta_1 X_t + u_t.$$

The omitted variables will, in this case, lead to substantial first-order autocorrelation in the error terms. The standard fix is to apply GLS assuming the following error structure

$$u_t = \rho u_{t-1} + \epsilon_t,$$

which, as we saw, will lead us to estimate the following equation:

$$(1-\rho L)Y_t = (1-\rho)\beta_0 + (1-\rho L)\beta_1 X_t + \epsilon_t.$$

With some rearranging, you will see that the "true" model can be rewritten as

$$(1-\gamma_4 L)Y_t = \gamma_1 + \gamma_2(1+\frac{\gamma_3}{\gamma_2}L)X_t + \varepsilon_t.$$

Applying the GLS correction, then, involves imposing the following "common factors" constraint:

$$\gamma_3 + \gamma_2\gamma_4 = 0.$$

Tests for common factors involve testing this constraint. Such considerations fit into the rubric of "general-to-specific modeling," which is treated in depth in Chapter 10.

8.5 An Aside: Newey–West Standard Errors

Newey–West standard errors are based on the same innovation that underpins the Huber–White correction for heteroskedasticity that has become virtually standard practice. Recall from earlier that with no heteroskedasticity or autocorrelation,

$$\begin{aligned}\text{Var}[b \,||\, X] &= E(X'X)^{-1}X'UU'X(X'X)^{-1} \,||\, X] \\ &= E(X'X)^{-1}X'\sigma^2 IX(X'X)^{-1} \,||\, X].\end{aligned}$$

But if either of these assumptions is violated, we write

$$\text{Var}[b \,||\, X] = E(X'X)^{-1}X'\sigma^2\Omega X(X'X)^{-1} \,||\, X].$$

The GLS estimator is

$$\hat{\beta} = (X'\Omega^{-1}X)^{-1}X'\Omega^{-1}y --$$

i.e., we need to invert the estimator of Ω, an nxn matrix.

The Huber–White correction is based on the observation that we don't really need the entire Ω matrix; what we really need is

$$X'\sigma^2\Omega X,$$

a *kxk* matrix. Newey–West standard errors are a simple extension of this logic, and are now easily computed in most statistical packages that include time series routines. There is a "lag length" associated with Newey–West estimators that refers *not* to the order of autoregressive process, but to the lag at which the autocorrelation is small enough to be ignored.[6]

Like GLS approaches, Newey–West estimation involves "fixing" the autocorrelation rather than thinking about its source and modeling it explicitly. While it is a more robust approach – unlike the Cochrane–Orcutt procedure, it includes no assumption about the nature of the autocorrelation in the error terms – it does not address omitted variable bias or dynamic . . . misspecification. We turn to these topics in Chapters 9 and 10.

8.6 Summary

- While students are often taught to " fix" autocorrelation by using Generalized Least Squares (GLS) or Newey-West methods, neither addresses the omitted variables bias that may be present due to autocorrelation driven by an omitted variable.
- While Durbin-Watson statistics (and to a lesser extent, Breusch-Godfrey tests) are commonly used to diagnose autocorrelation, they do not provide as much information as the autocorrelation function and Q-tests introduced in Chapters 4 and 5.
- Although coefficients estimated from GLS and OLS are the same in expectation, they can be dramatically different for a particular series.
- The GLS model is equivalent to an Autoregressive Distributed Log (ADL) model with particular restrictions on the coefficients. Testing the restriction is preferable to merely assuming it is correct. This is sometimes referred to as the "Common Factors (COMFAC) Critique," and will be addressed further in Chapter 9.

6 To understand this distinction, it may help to think about the autocorrelation function associated with an AR(1) process. While the order of the process is "1," the autocorrelation function exhibits decay; autocorrelations may be sizable for several lags.

References

Hendry D.F. and Mizon G.E. (1978). Serial Correlation as a Convenient Simplification, Not a Nuisance: A Comment on a Study of the Demand for Money by the Bank of England. *Economic Journal, Royal Economic Society* 88(351):549–63.

Mizon G.E. (1995). A simple message for autocorrelation-correctors: Don't. *Journal of Econometrics* 69:267–288.

9

Dynamic Specification: Distributed Lag Models

Chapter 8 introduced the COMFAC critique of the reflexive use of GLS to "correct" autocorrelation in a static regression model. We learned that the simple generalized least squares (GLS) model

$$\begin{aligned} &Y_t = \beta_0 + \beta_1 X_t + u_t \\ &u_t = \epsilon_t + \rho u_{t-1} \\ &-1 < \rho < 1 \\ &\epsilon_t \sim N(0, \sigma^2) \end{aligned} \tag{9.1}$$

is mathematically equivalent to this model

$$Y_t = \gamma_1 + \gamma_2 X_t + \gamma_3 X_{t-1} + \gamma_4 Y_{t-1} + \epsilon_t \tag{9.2}$$

with the constraint that $\gamma_3 + \gamma_2\gamma_4 = 0$ and we are advised to test the constraint. If we reject the null hypothesis, we infer that the (GLS) model is misspecified and we are advised to include lags of the dependent and independent variable in the model and permit free estimation of the associated coefficients.

This often raises red flags for well-trained analysts accustomed to working with cross-sectional data. Presumably, we would have included the lagged terms in our initial specification if there were any theoretical motivation. Moreover, adding them *ex post* typically results in a smaller estimate for the coefficient on our variable of interest.

Doing so, however, addresses a second, and equally important weakness of the GLS fix. Equation (9.1) seems to indicate that any change in the independent variable X_t is associated with only a contemporaneous change in the

Applied Time Series Analysis for the Social Sciences: Specification, Estimation, and Inference, First Edition. Regina M. Baker.

Companion website: www.wiley.com/go/baker

dependent variable Y_t – for example, that an increase in advertising in a given period is associated with an increase in sales only in that same period. By mechanically writing the standard regression model, we have failed to think about the questions addressed by dynamic specification. Does any effect of advertising on sales occur immediately, or is there a delay? If the latter, how long is the delay? Does any effect of advertising last more than one period? If so, how long?

This rich set of questions is addressed by "dynamic specification" – in the words of three LSE School luminaries, "appropriately matching the lag reactions of a postulated theoretical model to the autocorrelation structure of the associated observed time-series data." (Hendry, Pagan and Sargan 1984:1025). The answer to the analyst worried about whether the estimated coefficient in Equation (9.1) or (9.2) gives us the correct estimate of the marginal effect of X_t on Y_t, then, is that he is asking the wrong question; the dynamic relationship between Y_t and X_t is typically more interesting and more complicated than the single "marginal effect" number can convey. And to his first concern: the addition of a lagged dependent and/or independent variable need not be motivated by the belief that either is exerting any causal effect. Instead, it is an atheoretical means of modeling the autocorrelation that improves our ability to find the most appropriate dynamic specification.

This chapter introduces distributed lag and autoregressive distributed lag (ADL) models, the simplest regression-based models designed to assess dynamic effects. Definitions and further justification are provided in the next section. The Koyck model, originally seen in Chapter 1, is used to illustrate basic principles of dynamic specification and to introduce the "backward- and forward-thinking" logic that permits us to derive dynamic effects from the ADL equation. A subsequent section provides a typology of ADL(1,1) models. The chapter concludes with an introduction to methods for estimating the equations and testing restrictions.

9.1 Distributed Lag and Autoregressive Distributed Lag Models

Both types of model can be used to represent stationary series. "Distributed lag models" take the following general form:

$$Y_t = \alpha_0 + \beta_0 X_t + \beta_1 X_{t-1} + \beta_2 X_{t-2} + \cdots + \beta_k X_{t-k} + u_t, \quad k \in \{0,1,2\ldots\infty\}. \tag{9.3}$$

ADL models take the general form

$$\begin{aligned} Y_t = \alpha_0 + \alpha_1 Y_{t-1} + \alpha_2 Y_{t-2} + \cdots + \alpha_m Y_{t-m} + \beta_0 X_t + \\ \beta_1 X_{t-1} + \beta_2 X_{t-2} + \cdots + \beta_n Y_{t-n} + \epsilon_t \end{aligned} \quad (9.4)$$

where

- $m \in \{0,1,2\ldots\infty\}$,
- $n \in \{0,1,2\ldots\}$, and
- $\epsilon_t \sim N(0,\sigma^2)$.

ADL models are commonly denoted ADL(m,n), where m refers to the highest order AR term included and n refers to the longest lag on the X variable. In practice, it is rare to encounter models requiring $m > 2$. Because models with $m \leq 1$ and $n \leq 1$ subsume several models commonly seen in the economics literature, a typology is provided in a later section.

Notice that distributed lag models and autodistributed lag models are not mutually exclusive. For example, the standard regression model with a well-behaved error term

$$\begin{aligned} Y_t &= \alpha + \beta X_t + \epsilon_t \\ \epsilon_t &\sim N(0,\sigma^2) \end{aligned}$$

is technically both. Also notice that ADL models include white noise error terms by definition (Hendry, Sargan and Pagan 1984:1041). As a consequence, all but the COMFAC model can be estimated with OLS.

ADL models are a special case of vector autoregression (VAR) models, and like VAR, often elicit initial confusion and/or horror due to misunderstandings and misinterpretation. Three features require special attention: the motivation behind any lagged dependent variables (LDVs), the tension between theory and empirics, and the shift in interpretation from the marginal effects approach used with cross-sectional data to the dynamic effects approach appropriate for time series data.

First, the inclusion of LDVs need not be motivated by assumptions about causality any more than the inclusion of AR terms in ARIMA models was motivated by beliefs about causality. Instead, it is a strategy for modeling the autocorrelation that mitigates the danger of omitted variables bias. Recall from Section 1.2.2 that autocorrelation can be caused by omitting variables whose impact on the dependent variable persists over time, and if these omitted variables are also correlated with included variables, regression estimates will be biased. Thinking of autocorrelation as a nuisance to be swept into an error term "fix" not only fails to address the omitted variables bias, but misses the unique opportunity that longitudinal data afford. Since most of the variables we work with are strongly correlated with lags of themselves, lags of the dependent variable include information about variables that have been excluded even if the excluded variables are unmeasurable! Indeed, atheoretical ARIMA models outperform well-considered theoretical specifications so frequently in large part because lagged variables are such good proxies for omitted variables – in particular, for the correct set of omitted variables properly weighted to reflect their impact.

At the same time, the horror that many trained in regression approaches express when first encountering ARIMA, ADL, and VAR models is not entirely misplaced. It seems incredible to suggest that a model appropriately informed by theory could be inferior to one that ignores it entirely. David Hendry explained the evolution of his thought regarding the balance between theory and empirics in the preamble he wrote for the reprinted version of the "Dynamic Specification" chapter that he, Adrian Pagan and J. Denis Sargan had originally published in the *Handbook of Econometrics*. (D.F. Hendry, A.R. Pagan and J.D. Sargan, 1984; and D.F. Hendry, 1993)

> I realized at last what the essential difficulty was with my earlier empirical studies: the equation or model to be estimated was obtained from a theoretical derivation and was made stochastic by adding on an appropriately behaved 'error term.' The distributions of the dependent variables were then derived from those of the error terms and that generated the likelihood function for estimation, inference, etc. It all seemed coherent. Implicitly, however, it still assumed omniscience. In a theory framework, the model obtained is a function of the assumptions made and as such is freely created by its proprietor. In an empirical framework, the *data* are given, and so the distributions of the dependent variables are already fixed by what the data generation process created them to be – I knew that from Monte Carlo. But I kept missing the obvious point until Jean-Francois [Richard] drove it home during my CORE visit: the consequence of given data and a given theory model is that *the error is the derived component*, and one cannot make 'separate' assumptions about its properties. My model derivations were in essence back to front, and I had to reverse the process and obtain the error's distribution from that of the data distributions in empirical modelling... (Hendry 1993:73)

The critical issue, then, is achieving the correct balance between theory and empirics. In the same way that we would drop the assumption of a linear relationship if a scatterplot indicated curvilinearity, we must not let theory dictate a time series specification that's inconsistent with the observed autocorrelation structure.

Third, it is rarely sufficient to interpret a single regression coefficient as "the marginal effect" of X on Y. Instead, estimated parameters are used jointly to examine how the time path of our dependent variable changes in response to a change in the independent variable(s) of interest, in the same way that they were used in transfer function analysis in the ARIMA context. Losing the single-coefficient marginal effects interpretation may be less troubling if you remember the presidential popularity model from Chapter 1. We saw that the model could be written in mathematically equivalent ways that represent quite different causal stories. In the first, presidential popularity appeared to be driven by presidential popularity in the previous period and the change in disposable income from the previous to the current period. In the second, presidential popularity was driven by the history of changes in disposable income; presidential popularity in the previous period did not explicitly appear in the model at all.

The presidential popularity model was an example of a Koyck model, and in the next section, we will examine in it in detail. Because it does not include a white noise error term, it is not an ADL model, but its distributed lag form is useful for illustrating dynamic specification and forward- and backward-thinking. The subsequent section presents a general typology of ADL(1,1) models, listing their implied constraints and describing their implications for the time path of the dependent variable. Methods for estimating the equations and testing constraints will follow, and the chapter will conclude with a discussion of a more general form of distributed lag model, Almon Lag models.

9.2 The Koyck Model and Dynamic Specification

In Chapters 1 and 2, we examined the following two models of presidential approval over time (with slightly different notation)[1] and showed that they are mathematically equivalent:

$$\text{PresApp}_t = \beta_0(1-\lambda) + \lambda\text{PresApp}_{t-1} + \beta_1\Delta\text{Income}_t + \epsilon_t - \lambda\epsilon_{t-1} \tag{9.5a}$$

$$\begin{aligned}\text{PresApp}_t &= \beta_0 + \beta_1\Delta\text{Income}_t + \lambda\beta_1\Delta\text{Income}_{t-1} + \lambda^2\beta_1\Delta\text{Income}_{t-2} \\ &\quad + \lambda^3\beta_1\Delta\text{Income}_{t-3}\ldots + \epsilon_t \\ &= \beta_0 + \sum_{k=0}^{\infty}\lambda^k\beta_1\Delta\text{Income}_{t-k} + \epsilon_t.\end{aligned} \tag{9.5b}$$

1 Unfortunately, the notation tends to differ depending on context. Chapters 1 and 2 used standard regression notation; the notation here is standard for the Koyck model.

9.2.1 Motivation

Collectively, they are called a "Koyck" model, and are motivated by the following problem. Suppose you have some vague sense that a president's popularity today is affected not only by changes in income in the current period, but also by the changes in income for the last few periods:

$$\begin{aligned}\text{PresPop}_t &= \alpha + \beta_0 \Delta\text{Income}_t + \beta_1 \Delta\text{Income}_{t-1} + \beta_2 \Delta\text{Income}_{t-2} \\ &\quad + \beta_3 \Delta\text{Income}_{t-3} + u_t\end{aligned}$$

where $u_t \sim N(0,\sigma^2)$. It is not unreasonable to guess that recent changes in income have greater effect than changes further in the past – i.e., $\beta_0 > \beta_1 > \beta_2 > \beta_3$.

As long as the standard Gauss–Markov assumptions are met, there is nothing inherently wrong with using OLS to estimate this equation. In practice, however, the $\Delta\text{Income}_{t-k}$'s (lagged independent variables) are likely to be highly correlated, making it very difficult to obtain precise estimates. The Koyck model circumvents this problem by introducing an assumption about the relationship between the coefficients. In particular, it assumes that the coefficients on the lagged independent variables decline geometrically. The general form can be written as follows:

$$Y_t = \alpha + \beta_0\lambda^0 X_{t-0} + \beta_0\lambda^1 X_{t-1} + \beta_0\lambda^2 X_{t-2} + \beta_0\lambda^3 X_{t-3} + \cdots + \epsilon_t \tag{9.6}$$

where $-1 < \lambda < 1$ and $\epsilon_t \sim N(0,\sigma^2)$.

Notice that the model no longer requires estimation of several βk's. In their place, we need estimate only β_0 and λ. Doing so is facilitated by writing Equation (9.5b) as Equation (9.5a):

$$\begin{aligned}Y_t &= \alpha + \beta_0\lambda^0 X_{t-0} + \beta_0\lambda^1 X_{t-1} + \beta_0\lambda^2 X_{t-2} + \beta_0\lambda^3 X_{t-3} + \cdots + \epsilon_t \\ &= \alpha + \beta_0(1 + \lambda L + \lambda^2 L^2 + \lambda^3 L^3 + \cdots)X_t + \epsilon_t \\ Y_t &= \alpha + \frac{\beta_0}{1-\lambda L} X_t + \epsilon_t.\end{aligned}$$

Multiplying both sides by $1-\lambda L$ yields the following equation:

$$Y_t = (1-\lambda)\alpha + \lambda Y_{t-1} + \beta_0 X_t + \epsilon_t - \lambda\epsilon_{t-1}.$$

That is, our original model with geometrically declining coefficients on the lagged independent variable is mathematically equivalent to a model with a lagged dependent variable, only the contemporaneous independent variable, and an MA(1) error term. λ and β_0 are readily estimated, and the implied coefficients on Xt, X_{t-1}, X_{t-2}, etc., can be calculated as β_0, $\lambda\beta_0$, $\lambda^2\beta_0$, etc. respectively. The estimated λ gives us the rate of geometric decline.

For the sake of illustration, suppose $\lambda = 0.5$ and $\beta_0 = 3$. Then Equation (9.5b) becomes

$$\text{PresPop} = 50 + 3\Delta\text{Income}_t + 1.5\Delta\text{Income}_{t-1} + 0.75\Delta\text{Income}_{t-2} + \cdots + u_t.$$

Last period's change in income has half the impact on presidential popularity that the current period's does, and the period's before last has half again as much impact as last period's.

9.2.2 Marginal Effects, Dynamic Effects, and Backward- and Forward-thinking

Employing the marginal effects approach we are accustomed to using with cross-sectional data, we would infer that a one-unit change in X_t is associated with a change in Y_t equal to β_0. This represents only the contemporaneous (or short-term) effect, though. Those who studied ARIMA transfer functions carefully will recognize that by doing so we are taking too narrow a view of "the effect of X on Y." A complete accounting will incorporate longer-term effects as well. Indeed, we can apply the same process to Equation (9.5b) that we used to explain transfer functions and derive the long-term effects of X on Y.

To begin, suppose there is a change in X_t in period t. We can use Equation (9.5b) to write the equations representing the value of Y the first three periods following the change:

$$\begin{aligned}\text{PresApp}_{t+1} = \beta_0 + \beta_1\Delta\text{Income}_{t+1} + \lambda\beta_1\Delta\text{Income}_t + \lambda^2\beta_1\Delta\text{Income}_{t-1} \\ + \lambda^3\beta_1\Delta\text{Income}_{t-2} \ldots + \epsilon_t\end{aligned} \tag{9.8a}$$

$$\begin{aligned}\text{PresApp}_{t+2} = \beta_0 + \beta_1\Delta\text{Income}_{t+2} + \lambda\beta_1\Delta\text{Income}_{t+1} + \lambda^2\beta_1\Delta\text{Income}_t \\ + \lambda^3\beta_1\Delta\text{Income}_{t-1} \ldots + \epsilon_t\end{aligned} \tag{9.8b}$$

$$\begin{aligned}\text{PresApp}_{t+3} = \beta_0 + \beta_1\Delta\text{Income}_{t+3} + \lambda\beta_1\Delta\text{Income}_{t+2} + \lambda^2\beta_1\Delta\text{Income}_{t+1} \\ + \lambda^3\beta_1\Delta\text{Income}_t \ldots + \epsilon_t.\end{aligned} \tag{9.8c}$$

If you are familiar with calculus, you will quickly recognize that tracing the effect of the change in ΔIncome_t over time becomes a simple matter of taking the derivative of Equations (9.8a–9.8c) with respect to ΔIncome_t. Alternatively, you can think in terms of using the three equations to assess the marginal effect of ΔIncome_t on Y_{t+1}, Y_{t+2}, and Y_{t+3}, respectively.

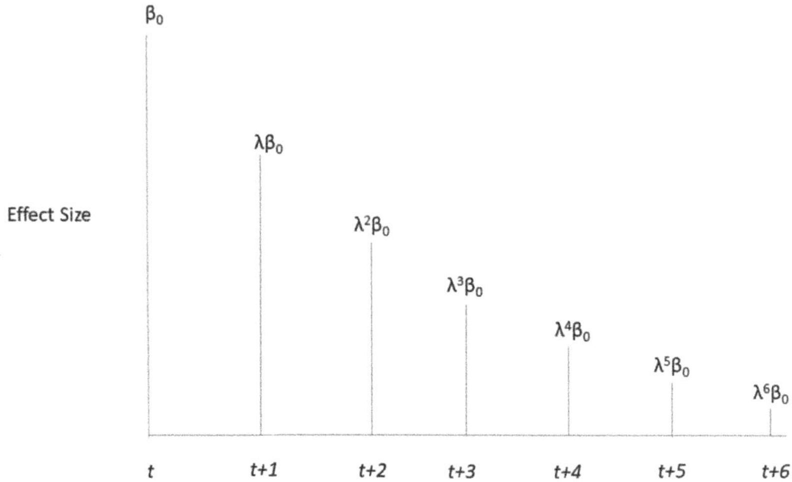

Figure 9.1 Koyck Model: effects of a one-unit-change in X_t on Y_t over time.

- Based on Equation (9.8a), each one-unit change in X_t will be associated with a $\lambda\beta_1$-unit change in Y_{t+1}, i.e., presidential approval in the first period following the change.
- Based on Equation (9.8b), each one-unit change in X_t will be associated with a $\lambda^2\beta_1$-unit change in Y_{t+2}, presidential approval in the second period following the change.
- Based on Equation (9.8c), each one-unit change in *Xt* will be associated with a $\lambda^3\beta_1$-unit change in Y_{t+1}, i.e., presidential approval in the third period following the change.

A pattern is becoming clear; in the *k*th period following the change in the independent variable, the impact on the dependent variable will be $\lambda^k\beta_1$ per unit. This relationship is shown in Figure 9.1.

The long-run effect of *X* on *Y*, then, is not is not an individual coefficient, but rather, the sum of effects over all periods. As in the case of our first-order transfer function, the total effect of *X* on *Y* can be derived as

$$\begin{aligned}
&\beta_0 + \lambda\beta_0 + \lambda^2\beta_0 + \lambda^3\beta_0 \cdots \\
&= \beta_0(1 + \lambda + \lambda^2 + \lambda^3 + \cdots) \\
&= \beta_0 \sum_{k=0}^{\infty} \lambda^k \\
&= \frac{\beta_0}{1-\lambda}.
\end{aligned}$$

While the mathematical derivation of the impact of a change in X on Y over time should be convincing, it wasn't necessary; the logic of "forward- and backward-thinking" is sufficient (but typically far from obvious). Compare the period-by-period impacts shown in Figure 9.1 with the coefficients in Equation (9.5b), and notice the following.

- The impact of a one-unit change in X_t on Y_{t+1} is equal to the coefficient on X_{t-1}.
- The impact of a one-unit change in X_t on Y_{t+2} is equal to the coefficient on X_{t-2}.
- The impact of a one-unit change in X_t on Y_{t+3} is equal to the coefficient on X_{t-3}.

And the pattern continues. Forward- and backward-thinking involves the recognition that if "a one-unit change in $X_t - 3$ is associated with a γ change in Y_t," for example, "A one-unit change in Xt must be associated with a γ change in Y_{t+3}." Both must be true in the same way that if I am ten feet ahead of you, you must be ten feet behind me. With practice, this logic will become second nature and help a great deal with intuition about time series dynamics.

In addition to assessing both the immediate and long-run effect of changes in X on Y, it is interesting to think about the time required for the long-run effect to be realized.

- The *median lag* is defined as the time required for the first half of the total change in Y to take place. In the case of the Koyck model, it is calculated as $-\frac{ln2}{ln\lambda}$.
- The *mean lag* is less intuitive, but defined as the average time required for the effect change in Y due to a change in X to be realized, and becomes the weighted average of all lags present in the model. For the Koyck model, it is computed as $\frac{\lambda}{1-\lambda}$.

9.3 General-to-Specific Modeling and the ADL(1,1) Model

We have seen that the GLS model used to correct for first-order autocorrelation is a special case of an ADL(1,1) model, which we can write

$$Y_t = \alpha_0 + \alpha_1 Y_{t-1} + \beta_0 X_t + \beta_1 X_{t-1} + \epsilon_t,$$

and that the Koyck model can be used to represent a relationship in which the effect of X on Y is geometrically declining over time. The motivation for general-to-specific modeling may be becoming clear. While researchers are often trained to write and estimate a regression model suggested by theory, doing so imposes restrictions that may not be true. The implied nature of the relationship between Y and X over time may not be a match for the autocorrelation structure of the series. Since several common economic models are encompassed by the ADL(1,1) model, estimating this general model and then testing the restrictions implied by theory is a sensible alternative. A brief catalogue is presented in Table 9.1.

In all of these cases, the ADL(1,1) model is an encompassing model that can be used to represent multiple theories. Estimating the ADL(1,1) model and then testing the restrictions our theory implies keeps us from making an assumption that is not true, and/or making assumptions about the impact of X on Y over time that are inconsistent with our data. Indeed, while we may have thought that writing a model based on theory was an appropriate theory-driven strategy, many would use the term "theory-laden": we do inferior work by making an assumption that may not be true. These ideas will be explored further in Chapter 11, where a general-to-specific modeling strategy is used in LSE/Hendry school fashion, and in the final chapters covering vector autoregression.

A few notes about the models are in order. First, notice that by writing the equations in ADL(1,1) form, we expose important differences between Model (1), the static regression model, and Model (2), GLS with an AR(1) error term. Model (3), the equation in first differences, is a commonly recommended strategy for working with non-stationary series, but we will discuss its limitations in Chapter 10. Model (4), the pure AR(1) process, is typically an excellent fit for the data. To the extent that it is a better fit than a model including important explanatory variables, though, it's often a better fit due to dynamic misspecification. The implied dynamics for Models (5)–(7) are easily derived, and will be explored further in the Exercises.

Finally, error correction models (ECMs) have become strongly identified with models involving cointegrated non-stationary series, covered in the next chapter, but they predate cointegration in the literature by decades – see Phillips (1954) and Sargan (1964), for example. Moreover, the basic principles are the same. ECMs are constructed in a way that emphasizes a long-run relationship between variables, the immediate impact of a disturbance in the relationship, and the speed of adjustment when the long-run relationship is disturbed. In a canonical paper, Davidson, Hendry, Srba, and Yeo incorporated an ECM in their model of the relationship between consumption (C_t) and income (Y_t). A simplified illustration begins by supposing both series are stationary (perhaps because they are measured in constant dollars), and the marginal propensity to consume – the fraction of consumers' income that they spend, rather than save – is constant over time. Then from economic theory,

Table 9.1 Common dynamic models and their relationship to the ADL(1,1) model.

Model	Equation	Constraint
1. Static regression	$Y_t = \alpha_0 + \beta_0 X_t + \varepsilon_t$	$\alpha_1 = \beta_1 = 0$
2. Common factor	$Y_t = \alpha_0 + \alpha_1 Y_{t-1} + \beta_0 X_t - \alpha_1 \beta_0 X_{t-1} + \varepsilon_t$	$\beta_1 = -\alpha_1 \beta_0$
3. Equation in first differences	$\Delta Y_t = \alpha_0 + \beta_0 \Delta X_t + \varepsilon_t$	$\alpha_1 = 1,\ \beta_0 = -\beta_1$
4. First-order AR process	$Y_t = \alpha_0 + \alpha_1 Y_{t-1} + \varepsilon_t$	$\beta_0 = \beta_1 = 0$
5. Leading indicator equation	$Y_t = \alpha_0 + \beta_1 X_{t-1} + \varepsilon_t$	$\alpha_1 = \beta_0 = 0$
6. First-order finite distance lag	$Y_t = \alpha_0 + \beta_0 X_t + \beta_1 X_{t-1} + \varepsilon_t$	$\alpha_1 = 0$
7. Dead start model	$Y_t = \alpha_0 + \alpha_1 Y_{t-1} + \beta_1 X_{t-1} + \varepsilon_t$	$\beta_0 = 0$
8. Partial adj. equation	$Y_t = \alpha_0 + \alpha_1 Y_{t-1} + \beta_0 X_t + \varepsilon_t$	$\beta_1 = 0$
9. ECM	$\Delta Y_t = \alpha_0 + \beta_0 \Delta X_t + (\alpha_1 - 1)(Y_{t-1} - X_{t-1}) + \varepsilon_t$	$\alpha_1 - 1 = -(\beta_0 + \beta_1)$

Adapted from Hendry, Pagan and Sargan 1984:1042.

$$C_t = cY_t$$

represents the long-term relationship between consumption and income, where c is the fraction of national income that is spent rather than saved. In the short-term, though, events can push this relationship off its equilibrium. For example, suppose the government distributes stimulus checks, temporarily raising income, and consumers do not spend the windfall immediately. Then $C_t < cY_t$ for that period. An ECM can be written as

$$\Delta C_t = \alpha_0 + \beta_0 \Delta Y_t + (\alpha_1 - 1)(C_t - cY_t) + \epsilon_t \tag{9.9}$$

where $\epsilon \sim N(0, \sigma^2)$.

- β_0 provides a way to measure the short-term effect of the change in income – i.e., the same-period change in consumption associated with a one-unit change in income.
- We know that in the long run, $C_t = cY_t$. Then the expression $C_t - cY_t$ in Equation (9.9) represents the extent of the disequilibrium.
- $\alpha_1 - 1$ indicates the speed of adjustment back to the long-run equilibrium.

A detailed example is provided in Chapter 10.

9.4 Estimating Models with Lagged Dependent Variables

Another common source of confusion for those new to ADL models concerns their estimation. We will see below that if the error term in a model with an LDV is autocorrelated, the estimated coefficient is not only biased, but inconsistent; the bias does not become smaller with a longer time series. By definition, though, the error term in ADL models is a white noise error term. Remember that the motivation behind including lags of the dependent and/or independent variable is to model the autocorrelation included variables rather than leaving it in the error term; doing so facilitates assessment of the dynamic specification and mitigates the potential for omitted variables bias in cases where the autocorrelation is driven by omitted variables that are correlated with included variables. If initial estimation of an ADL model indicates that there is autocorrelation remaining, additional lags of the dependent and/or independent variable should be added to the model to cover the autocorrelation.

9.4.1 LDV with Uncorrelated Errors

For simplicity's sake, consider a stationary model with only a lagged dependent variable:

$$Y_t = \alpha + \beta Y_{t-1} + u_t, \tag{9.10}$$

where $-1 < \beta < 1$ and $u_t \sim (0, \sigma^2)$.

We can rewrite the equation as follows:

$$\begin{aligned} (1-\beta L)Y_t &= \alpha + u_1 \\ Y_t &= \frac{\alpha}{1-\beta} + \frac{u_t}{1-\beta L} \\ Y_t &= \frac{\alpha}{1-\beta} + u_t + \beta u_{t-1} + \beta^2 u_{t-2} + \beta^3 u_{t-3} + \cdots + \beta^{t-1} u_1. \end{aligned} \tag{9.11}$$

To evaluate the correlation between the included variable Y_{t-1} and the error terms u_{t-k}, we multiply Equation (9.11) by the u_{t-k} of interest and take expectations. That is,

- First multiply Equation (9.11) through by u_t and take expectations.
- Then multiply the equation through by u_{t-1} and take expectations.
- Do the same for u_{t-2}.
- etc.

You will see that

$$\begin{aligned} E(Y_t u_t) &= \sigma^2, \\ E(Y_t u_{t-1}) &= \beta\sigma^2, \\ E(Y_t u_{t-2}) &= \beta^2\sigma^2, \\ &\vdots \\ E(Y_t u_{t-k} &= \beta^k\sigma^2. \end{aligned}$$

That is, Y_t is correlated with the current and previous disturbances, but uncorrelated with all future disturbances. Ergo Y_{t-1} is *not* correlated with u_t, but *is* correlated with u_{t-1} and its predecessors.

Consequently, OLS estimates of β are biased in finite samples. The bias becomes smaller, however, as the sample size becomes larger – i.e., OLS estimates are consistent in this case.

9.4.2 LDV with Autocorrelated Errors

Suppose that instead of being independently distributed, the u_t in Equation (9.11) follows an AR(1) process:

$$u_t = \rho u_{t-1} + \epsilon_t,$$

where $\epsilon_t \sim N.i.i.d(0, \sigma_\epsilon^2)$.

Then

$$\begin{aligned} E(Y_{t-1}u_t) &= E[Y_{t-1}(\rho u_{t-1} + \epsilon_t)] \\ &= \rho E(Y_{t-1}u_{t-1}) + E(Y_{t-1}\epsilon_t) \\ &= \rho E[(\alpha + \beta Y_{t-2} + u_{t-1})u_{t-1}] \\ &= \rho\beta E(Y_{t-1}u_t) + \rho\sigma_u^2 \\ (1-\rho\beta)E(Y_{t-1}u_t) &= \rho\sigma_u^2 \\ E(Y_{t-1}u_t) &= \frac{\rho\sigma_u^2}{1-\rho\beta} \end{aligned}$$

To express this in terms of σ_ϵ^2, derive

$$E(u_t^2) = \frac{\sigma_\epsilon^2}{1-\rho^2}$$

and substitute:

$$E(Y_{t-1}u_t) = \frac{\rho\sigma_\epsilon^2}{(1-\rho\beta)(1-\rho^2)}.$$

Therefore, OLS estimates are biased, and worse, they are inconsistent – the bias doesn't decrease as the sample size becomes bigger. The model must be estimated with maximum likelihood or instrumental variables methods.

9.5 Testing Constraints

Methods for testing constraints will be familiar from earlier studies. There are two general approaches. First, the F-test for linear restrictions can be used to test the null hypothesis that all restrictions hold. Let SSE_r be the error sum of squares from the restricted model, and SSE_u be the error sum of squares from the general ADL(1,1) model. Then the test statistic becomes

$$F = \frac{(\text{SSE}_r - \text{SSE}_u)/m}{\text{SSE}_u/(T-k)}$$

where

- m is the number of constraints,
- T is the number of time points, and
- k is the number of estimated parameters.

The test statistic is compared to the F-distribution with $(m, T-k)$ degrees of freedom. The null hypothesis is that the restrictions are true. Rejecting the null hypothesis, then, indicates that the proposed model is misspecified.

Alternatively, constraints can be tested separately using the familiar Wald, likelihood ratio, or Lagrange multiplier tests.

References

Davidson, J.E.H., Hendry D.F., Srba F., and Yeo S. (1978). Econometric Modelling of the Aggregate Time-Series Relationship Between Consumers' Expenditure and Income in the United Kingdom. *Economic Journal* 88:661–692.

De Boef, S. and Keele, L. (2008). Taking Time Seriously. *American Journal of Political Science* 52, 1 (Jan.): 184–200.

Hendry, D.F. (1993). Econometrics: Alchemy or Science? Essays in Econometric Methodology. Oxford, UK and Cambridge, US: Blackwell.

Hendry, D.F., Pagan, A.R., and Sargan, J.D. (1984). Dynamic specification. In *Handbook of Econometrics* (eds Z. Griliches and M.D. Intriligator), vol. II. Amsterdam: North Holland, ch. 18.

Phillips, A.W. (1954). Stabilization Policy in a Closed Economy. *Economic Journal* 64:290–323.

Sargan, J.D. (1964). Wages and Prices in the United Kingdom: a Study in Econometric Methodology. In *Econometric Analysis for National Economic Planning* (eds P.E. Hart, G. Mills and J.N. Whittaker), London: Butterworths.

10

Regression with Non-stationary Series: Cointegration and Error Correction Models

The coverage of ADL models in Chapter 9 emphasized three features common in more recent time series approaches. First, general-to-specific modeling provides a means of testing restrictions implied by theory, instead of merely assuming they are correct, and will therefore lead to better model specification. Second, modeling autocorrelation by adding lags of the dependent and/or independent variables mitigates the omitted variables bias that is introduced when autocorrelation driven by omitted variables is instead modeled in the error component of the equation. As a consequence, though, estimated coefficients do not typically represent marginal effects, so third, the catalogue of ADL(1,1) variations facilitates recognition of the dynamic relationship between the two variables – e.g., how the dependent variable changes over time in response to changes in the independent variable.

This chapter addresses the problem of spurious regressions that plagues associations involving non-stationary series. In the case of ARIMA models, differencing non-stationary series was a necessary first step, and working with differenced series is also an option in the ordinary regression context. Doing so, however, narrows the focus to short-term relationships between variables and eliminates the possibility of examining their long-term relationship. This problem provided the motivation for the work on cointegration that earned Clive Granger the Nobel Prize in 2003.[1] Some intuition is conveyed in Figure 10.1.

1 He shared the prize that year with Robert Engle, whose award was based on his development of Autoregressive Conditional Heteroscedasticity (ARCH) models. Although Engle and Granger coauthored the 1987 article formally introducing cointegration and ECM for non-stationary series, the groundwork was laid in a 1983 discussion paper.

Applied Time Series Analysis for the Social Sciences: Specification, Estimation, and Inference, First Edition. Regina M. Baker.

Companion website: www.wiley.com/go/baker

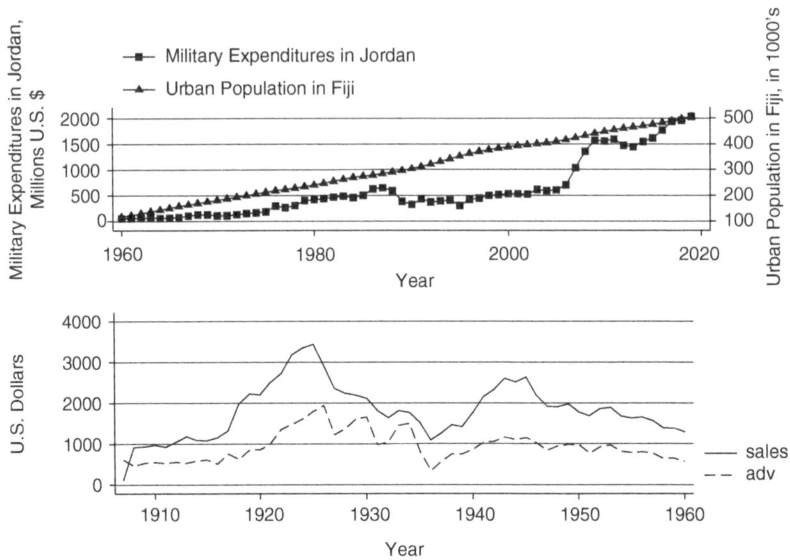

Figure 10.1 A Spurious Relationship vs. Cointegrated Series.

The series in the top panel are both increasing and highly correlated. Although it's a nonsense regression, the OLS results presented in Chapter 3 indicate that the size of the urban population in Fiji explained 74% of the variance in Jordan's military expenditures between 1960 and 2019. These two series, though, do not appear to be moving together in the same way that the two series in the bottom panel are. The shapes of the paths that Lydia Pinkham's sales and advertising series take are approximately the same; like a figure skating pair, they do not appear to drift very far apart before moving together again. Michael Murray described this tendency, alternatively, as being like those of a drunk and her dog:

> The drunk sets out from the bar, about to wander aimlessly in random-walk fashion. But periodically she intones "Oliver, where are you?", and Oliver interrupts his aimless wandering to bark. He hears her; she hears him. He thinks, "Oh, I can't let her get too far off; she'll lock me out." She thinks, "Oh, I can't let him get too far off; he'll wake me up in the middle of the night with his barking." Each assesses how far away the other is and moves to partially close that gap. (Murray, 1994:37)

In this chapter we will see that the Lydia Pinkham sales and advertising series are *cointegrated* but the Jordanian military expenditures and Fijian urban population series are not. Cointegrated series have three special properties. First, by the definition of cointegration, the residual from regressing one on the other is stationary. As such, it meets a minimal requirement for estimation by OLS that is not met by non-stationary series that are not cointegrated. Second, the estimated coefficient not only converges to a limiting distribution as the number of usable time points increases, regression estimates from cointegrated series are super-consistent. That is, the OLS estimates will converge to their true values faster than they do for stationary series. Finally, they can always be represented by the ECM model we examined in Chapter 9, which allows us to evaluate the long- and short-run relationship between the series.

The next section expands upon the reasons for the false positive results obtained when using OLS to estimate regressions of one non-stationary variable on another. The definition of cointegration follows, and then tests for cointegration are introduced and applied to the two sets of series depicted in Figure 10.1. The second half of the chapter revisits the ECM in the context of non-stationary series and provides two examples using the Engle-Granger method.

10.1 Non-stationary Series and Spurious Regression

Finding statistically significant nonsense regressions of non-stationary series is child's play. In addition to the regression of Jordan's military expenditures on the urban population of Fiji presented earlier, consider the results of three estimations of the following OLS regression, presented in Table 10.1:

$$Y_t = \alpha + \beta X_t + a_t$$

where *at* is assumed (incorrectly, in these cases) to be independently and normally distributed with mean 0 and variance σ^2.

The results suggest that every additional megawatt-per-capita of electricity consumption in France is associated with a population increase of 800,000 men in Ecuador, that each additional year of life expectancy in Greece is associated with \$560 million in additional government spending in Kenya, and that an increase of \$1000 in Norwegian GDP per capita is associated with a reduction in Japanese infant mortality of 150 per 100,000 live births. In all three cases, the results are statistically significant at the $p < 0.001$ level, and the associated R^2's range from 0.47 to 0.93. All three defy credulity, however, and share the additional characteristic, common to spurious relationships between non-stationary series, of Durbin-Watson statistics very close to zero, indicating strong autocorrelation in the residuals.

In Chapter 3, we noted an obvious reason for correlations between trending series: two series moving in the same direction will necessarily be positively correlated, and two series moving in opposite directions will be necessarily negatively correlated. There is an additional problem, however, in the case of unit root series, even those that are mean-stationary. Unless there is a cointegrating relationship between the variables in question, the standard errors from OLS will be grossly underestimated, leading to unwarranted confidence in the statistical significance of the coefficients.

Granger and Newbold demonstrated this problem by examining simulations of randomly generated unit root processes. Although the series were independent by construction, they found that "using the traditional *t* test at the 5% level, the null hypothesis of no relationship between the two series would be rejected (wrongly) on approximately three-quarters of all occasions." (Granger and Newbold, 1974: 115). Warning signs that the statistically significant result does not indicate a true relationship include the combination of a high R^2 and a very low DW statistic, as we observed in the estimations of the spurious regressions presented in Chapter 3 and in Table 10.1. In later work, Phillips demonstrated that "in the Granger-Newbold regressions of independent random walks the usual t-ratio significance test does not possess a limiting distribution but actually diverges as the sample size $T \to \infty$." (Phillips, 1985:2) As a result, the bias

Table 10.1 More Spurious Regressions of Non-stationary Series.

Y_t	X_t	$\hat{\alpha}$	$\hat{\beta}$	R^2	DW Statistic
Male Population	Electricity Consumption	0.84	0.80	0.93	0.085
of Ecuador (Millions)	in France (MW per cap)	(< 0.001)	(<0.001)		
Government Consumption	Life Expectancy	−40	0.56	0.63	0.051
in Kenya (Billions U.S.$)	in Greece, Years	(<0.001)	(<0.001)		
Infant Mortality	GDP per capita	13	−0.15	0.47	0.048
in Japan, per 1000 births	in Norway, Thousands U.S.	(<0.001)	(<0.001)		

P-values are in parentheses below estimates. Data Source: *World Development Indicators*, downloaded from https://datatopics.worldbank.org/world-development-indicators/ Dec 12, 2020. The data fields used were SP.POP.TOTL.MA.IN,EG.USE.ELEC.KH.PC, NE.CON.GOVT.CD, SP.DYN.LE00.IN, SP.DYN.IMRT.IN, and NY.GDP.PCAP.CD.

toward incorrectly rejecting the null hypothesis of no relationship *increases* as T gets larger.

Fortunately, approaches covered in previous chapters will address the estimation problem. First, adding lagged values of both the independent and dependent variables – i.e., estimating an ADL model – addresses many of the estimation problems, but some hypothesis tests will require the use of special distributions. As we saw in Chapter 9, this will also change the interpretation of the coefficients. Second, the Cochrane-Orcutt model correcting for first-order autocorrelation, introduced in Chapter 8, is asymptotically equivalent to the model in first differences, and so will also address the estimation problem.[2] Third, as we saw in the case of ARIMA models, transforming the non-stationary series by differencing them is another solution. Finally, non-stationary series can be estimated as ECM models if they are cointegrated. The remainder of this chapter introduces the formal definition of cointegration, cointegration tests, and one method for estimating ECMs for cointegrated series. Keep in mind, though, that they solve only the estimation problem. Just as in the case of ordinary regression with cross-sectional data, association is no guarantee of causation and the possibility of spurious relationships must always be considered.

10.2 Cointegration and Cointegration Tests

The culprit behind the bad behavior of significance tests when regressing one non-stationary series on another is a non-stationary error term, so it may be no surprise that the definition of cointegration follows from properties of integrated series. We will use this insight, in fact, to distinguish between series that are "cointegrated," and series that are not. While cointegration does not guarantee that there is a causal relationship, regressions involving cointegrated variables will yield consistent estimates and will have stationary error terms. Cointegration is also a prerequisite for estimating ECM's for non-stationary series.

10.2.1 Integrated Series: Definitions and Properties

A series is said to be "Integrated of Order d", I(d), where d is an integer, if the series is non-stationary but can be made stationary by differencing d times. In

2 See Hamilton 1994:561–562 for further details regarding both of these approaches.

social science applications it is rare to encounter series that are integrated at higher orders than 2, so most series will fit into the following three categories.

- "Integrated of Order 1", I(1): A process that is non-stationary, but can be made stationary by differencing once.
- "Integrated of Order 2", I(2): A process that is non-stationary, but can be made stationary by differencing twice.
- An I(0) process is a stationary series.

The definition of cointegration will rely on several properties of integrated series. Let X_t, Y_t and Z_t be three time series, and let a and b be scalars (constant terms).

- If Y_t is I(d), then any linear transformation of Y_t – e.g., $a + bY_t$ – will also be I(d).
- If X_t and Y_t are both I(0), then any linear combination of X_t and Y_t – i.e., $Z_t = aX_t + bY_t$, – will also be I(0).
- If X_t is I(0) and Y_t is I(1), then any linear combination of the two will be I(1) – i.e., a linear combination of a stationary and non-stationary series is non-stationary.
- If X_t is I(d_1) and Y_t is I(d_2), then $Z_t = aX_t + bY_t$ is I(d_2), where $d_2 > d_1$. In other words, if two series are integrated of different orders, a linear combination of the series will be integrated of the higher order.
- if X_t and Y_t are both I(d), then linear combinations of the two – i.e., $Z_t = aX_t + bY_t$ – are *usually* I(d), but sometimes I(d*), where $d* < d$. That is, if two series are both integrated of the same order, linear combinations of the two are usually integrated of that order, but are sometimes integrated of a lower order.

10.2.2 Cointegration

This brings us to a definition of cointegration: two variables are *cointegrated* if there is a long-run equilibrium relationship between them. More formally, two integrated series X_t and Y_t are said to be cointegrated if they are integrated of the same order, and the linear combination $Z_t = Y_t - AX_t$ is integrated at a lower order. A is referred to as the "cointegrating parameter," and $[1 - A]$ (or any multiple) is the "cointegrating vector." Notice that the definition is agnostic about which variable is the dependent variable and which is the independent variable, if such a relationship is plausible.

In the examples later in the chapter, we will examine pairs of I(1) series, and test the hypothesis that a linear combination formed by the pair is I(0); that is, stationary.

It is also possible for a set of more than two variables to be cointegrated, as we will see below.

10.2.3 Testing for Cointegration

Engle and Granger (1987) introduced a straightforward method for testing for cointegration that exploits the fact that the error term from an ordinary regression can be written as a linear combination of the variables of interest. In the bivariate case, for example, where X_t and Y_t are the variables of interest, A is a scalar to be estimated, and e_t is the error term,

$$Y_t = \alpha + AX_t + e_t$$

can be rewritten as

$$e_t = Y_t - \alpha - AX_t \tag{10.1}$$

Notice that e_t is a linear combination of X_t and Y_t; it differs from the Z_t in our definition of cointegration only by the addition of the constant term α. A level shift in the series, though, does not affect its order of integration, so $\hat{e}_t$ can be used in cointegration tests as follows.

1. Determine that the two series are integrated of the same order using the methods described in Chapter 3. If there are more than two variables and they are not all integrated of the same order, at least two of them must be at the highest order of integration for the combination of variables to be cointegrated. For example, a set of three series that are I(2), I(1) and I(2) may be cointegrated, but a set of three series that are I(2), I(1) and I(1) cannot be.
2. Regress one series on the other(s), and save the estimated residuals, $\hat{e}_t$.
3. Evaluate the order of integration of the residuals.

   - The Cointegrating Regression Durbin-Watson (CRDW) test provides a quick-and-dirty test based on output that can often be easily generated with the regression results. The CRDW is calculated using the same formula as the ordinary DW statistic:

$$\text{CRDW} = \frac{\sum_{t=2}^{T}(\hat{e}_t - \hat{e}_{t-1})^2}{\sum_{t=1}^{T}\hat{e}_t^{\,2}} \tag{10.2}$$

   - Banerjee et. al. introduced an easy rule of thumb involving a comparison with estimated R^2. If CRDW $< R^2$, cointegration is unlikely. (Banerjee et. al., 1986)
   - Alternatively, an analogue to the ADF test presented in Chapter 3 can be employed. The correct tests do not include either a trend or constant term

in the estimating equation. If the estimated error term is $\hat{e}_t$, for example, the ADF estimating equation takes the form

$$\Delta\hat{e}_t = \alpha\,\hat{e}_{t-1} + a_t + \sum_{i=1}^{p}\gamma_i\Delta\hat{e}_{t-i}$$

– just as in the top right-most column of Table 3.2. Also, because we are testing the order of integration of an estimated series rather than an actual realization, we must use a special table of critical values, presented in the Appendix.

10.3 Testing for Cointegration: Two Examples

Earlier in the chapter we noted that Lydia Pinkham's sales and advertising series seemed to track each other, while the Jordanian military expenditures and Fijian urban population series did not. (See Figure 10.1.) Let's see how well our intuition regarding cointegration is reflected in Engle-Granger test results.

10.3.1 Example 1: Lydia Pinkham's Sales and Advertising Series

The first step – verifying that both series are integrated of the same order – has already been completed in Chapter 7. Both series exhibited unit roots, but could be made stationary by differencing once. Moreover, the correlograms of the differenced series did not indicate overdifferencing. Thus, we inferred that both are I(1) series.

The next step is to regress one series on the other and save the residuals. Since cointegration does not presume an order of causality, the estimation can be done using either sales or advertising as the dependent variable for this purpose. Indeed, it is often useful to test both ways, as the test statistics may differ substantially if the variances of the two series differ substantially. First, to facilitate comparison with the ARIMA model, sales is used as the dependent variable and the following model was estimated using OLS:

$$Sales_t = \underset{(142)}{451} + \underset{(.141)}{1.45}\ \text{Advertising}_t + \hat{u}_t \qquad (10.3)$$

where standard errors of the estimates are presented in parentheses.

To be cointegrated, $\hat{u}_t$ must be integrated at a lower order than the original series. In this case, since the original series were both I(1), testing for cointegration entails testing the hypothesis that $\hat{u}_t$ is I(0) – that is, stationary. Remember that in the case of cointegration tests, no constant term or trend term should be included in our ADF tests. We find that no lags are required (i.e., we can use the ordinary EG test instead of an Augmented EG test), and the estimated test statistic is –4.375.[3] Since there are two variables in our cointegration test, we compare this to first set of critical values presented in Appendix C. The test statistic is less than the critical value for the .01 significance level (–4.123), so we reject the null hypothesis that $\hat{u}_t$ has a unit root with $p < .01$.

Now let's perform the test using advertising as the dependent variable in the EG estimation. The estimated equation is

$$\text{Advertising}_t = \underset{(86.4)}{99.5} + \underset{(0.049)}{0.462}\,Sales + \hat{u}2_t \tag{10.4}$$

where standard errors of the estimates are presented in parentheses. In this case, also, we find that no lags are required, and the estimated test statistic is –4.187. While it is not identical to the test statistic when sales was used as the dependent variable in the EG test equation, it is also less than the critical value for the 0.01 significance level. We reject the null hyponthesis of a unit root in $\hat{u}2_t$, and infer that the sales and advertising series are cointegrated.

Later in the chapter, we will estimate ECM models for these series.

10.3.2 Example 2: Military Expenditures in Jordan and Urban Population in Fiji

Now let us what happens when we perform a cointegration test for one of our nonsense regressions, the regression of military expenditures in Jordan on the urban population in Fiji. The first step is to determine the order of integration of each series, which we can do using the ADF tests introduced in Chapter 3.

Preliminary ADF estimations for the series indicated that one lag was required for the military expenditure series and two were required for the urban population series. The results of these estimations are presented in the first and third rows of Table 10.2.

The estimated test statistic for the raw military expenditure series is –1.270, larger than the 10% critical value of –3.175, and the p-value for the MacKinnon test is 0.895; we cannot reject the null hypothesis of a unit root. The estimated test statistic for the differenced series is –4.672, smaller than the 1% critical value of –4.139, and the p-value for the MacKinnon test is 0.0007. We reject

3 A careful researcher may wonder if this warrants further investigation, since the coefficient on the fourth lag is sizeable and barely misses statistical significance, and our time series is not very long. We will do this later in the chapter by examining the residuals of the estimated ECM.

Table 10.2. Testing for Cointegration Between Miltary Expenditures in Jordan and the Urban Population in Fiji.

Series	α	θ_0	β	Test Statistic	1% Critical Value	5% Critical Value	10% Critical Value	MacKinnon P-value
Military Expenditure	−0.051	−18.8	2.27	−1.270	−4.132	−3.492	−3.175	0.8951
in Jordan	(0.040)	(23.5)	(1.24)					
Differenced	−0.983	−12.3	1.48	−4.672	−4.141	−3.496	−3.178	0.0461
Military Expenditure	(0.210)	(23.9)	(0.747)					
Urban Population	−0.053	6.72	0.358	−2.745	−4.135	−3.493	−3.176	0.2179
in Fiji	(0.019)	(1.99)	(0.130)					
Differenced	−0.322	2.00	−.004	−3.536	-4.141	−3.496	−3.178	0.0357
Urban Population	(0.091)	(0.617)	(0.006)					
$\hat{u}_{JFt}$	−0.053			−1.381	−4.124	−3.461	−3.130	N.A.
	(0.038)							
$\hat{u}_{FJt}$	−0.057			−1.810	−4.124	−3.461	−3.130	N.A.
	(0.031)							

Standard errors are reported in parentheses. In all cases, preliminary ADF estimations included 4 lags. Final estimations include one lag, three lags, two lags, four lags, one lag and one lag, respectively, leaving us with 58, 55, 57, 54, 58 and 58 cases, in order.

the null hypothesis of a unit root in the differenced series, suggesting that the raw series is not integrated at an order higher than 1. After verifying that the autocorrelation function does not indicate overdifferencing, we infer that the raw series is I(1).

The estimated test statistic for the raw urban population series is −2.745, larger than the larger than the 10% critical value of −3.176, and the p-value for the MacKinnon test is 0.218; we cannot reject the null hypothesis of a unit root. The estimated test statistic for the differenced population series is −3.536. Even with four lags, we are able to reject the null hypothesis of a unit root in the differenced series with $p < .05$. In this case also, the ACF does not indicate overdifferencing, so we infer that the raw series is I(1).

Having confirmed that the two series are integrated of the same order, we now estimate a regression with the two variables. The estimated equation is

$$MilExp_{Jordan} = \underset{(107)}{-688} + \underset{(.324)}{4.22} UrPop_{Fiji} + \hat{u}_{JFt} \qquad (10.5)$$

where standard errors are shown in parentheses. Now $\hat{u}_{JFt}$ can be used in our cointegration test. Remember that while the estimating equation is identical to the ADF estimating equation with no trend or constant term, we must use the critical values presented in Appendix C. In this case, the test statistic from the estimating equation with one lag is −1.381, higher than −3.130, the critical value at the .10 level of statistical significance. We cannot reject the null hypothesis of a unit root in the residuals of the cointegrating equation; we infer that the two series are not cointegrated.

The estimated equation when we reverse the dependent and independent variables is

$$UrPop_{Fiji} = \underset{(11.3)}{200} + \underset{(.014)}{.176}\, MilExp_{Jordan} + \hat{u}_{JFt} \tag{10.6}$$

where standard errors are shown in parentheses. The test statistic from the ADF estimating equation with one lag is −1.810, also higher than critical value at the 0.10 level of statistical significance. Again, although the test statistics differed, neither permits us to reject the null hypothesis of a unit root at the 0.10 level of statistical significance, indicating that these two series are not cointegrated.

10.4 Error Correction Models for Non-stationary Series

In Chapter 9, we saw that error correction models (ECMs) provide a convenient means of conveying the long-term and short-term relationships between two variables, and listed the restriction that must be tested in the case of stationary series.

In the case of non-stationary series, Engle and Granger have shown that cointegration is both a necessary (1991) and sufficient (1987) condition for series to be modeled in ECM form. The Granger Representation Theorem holds that if two variables X_t and Y_t are cointegrated, then the relationship between them can be expressed as an error correction mechanism:

$$\Delta Y_t = \alpha_0 + \alpha_1 \Delta X_t + \alpha_2 u_{t-1} + \epsilon_t, \tag{10.7}$$

where u_{t-1} is the lagged residual from the cointegrating equation (10.1) and $\epsilon_t \sim N(0,\sigma^2)$.

As we saw in the case of stationary variables,

- u_{t-1}, the error term from the cointegrating equation, is non-zero when X_t and Y_t are out of their long-run equilibrium.

- α_1 captures the short-term response by Y_t to a change in X_t. In other words, it gives a sense of how far out of equilibrium an innovation in X_t pushes Y_t.
- α_2, then, gives the rate of speed at which X_t and Y_t return to equilibrium.

10.4.1 Estimating the ECM

Engle and Granger provided a simple method for estimating ECMs that has remained popular despite critical limitations. After verifying that the series are cointegrated, as explained in section 10.2.3, save the estimated residuals from the cointegration equation (10.1), and estimate Equation (10.7).[4]

10.4.2 Example 1: Lydia Pinkham's Sales and Advertising

In Chapter 7, we used an ARIMA transfer function model to assess the impact of advertising spending on Lydia Pinkham's sales. Since both series were I(1), it was necessary to difference both series first; consequently, the model restricted our attention to the short-term relationship between sales and advertising. An ECM approach, on the other hand, will permit us to examine both the long- and short-term relationships.

Modeling Lydia Pinkham's sales and advertising data as a single-equation ECM provides a useful cautionary tale, however. Recall that we found evidence that sales might affect advertising, and indeed, companies' sales revenue is a typical consideration when deciding advertising budgets. Ignoring reciprocal effects exposes us to the problem of simultaneity bias – in this case, overestimating the effect of advertising on sales by failing to account for the effect of sales on advertising. In this section, we will proceed with single-equation simple ECM analyses anyway so that we can compare the results to VAR models in Chapters 12 and 13.

In this case, Equation (10.3) expresses the long-run equilibrium relationship between sales and advertising: each dollar spent on advertising was associated with 1.46 in sales – a substantial return. Having tested the residuals for nonstationarity and rejected the null hypothesis of a unit root, we believe we have trust the estimated statistical significance, $p < .001$.

Finally, we use the estimated residual from Equation (10.3) to estimate the ECM:

$$\Delta Sales_t = \underset{(30.2)}{22.7} + \underset{(0.152)}{.719}\ \Delta \text{Advertising}_t - \underset{(0.089)}{.211}\ \hat{u}_{t-1} + \hat{e}_t \tag{10.8}$$

4 I have shown the two-variable case here, but the same approach can be used with more than two variables.

where standard errors of the estimates are presented in parentheses. As in the non-stationary case introduced in Chapter 9, the estimated coefficients convey the contemporaneous effect of an increase in sales when the relationship is in equilibrium, and the speed with which the long-run equilibrium is restored after the equilibrium is disrupted.

- In equilibrium, $\hat{u}_{t-1} = 0$. $\alpha_1 = .719$ indicates that a dollar increase in advertising from time $t-1$ to time t is associated with a 71.9-cent increase in sales during the same period. Notice that such an increase would throw the relationship off its long-run equilibrium path; since sales increased substantially more than the 46.1 cents we would expect in long-run equilibrium.
- $\hat{\alpha}_2 = -.211$ indicates that 21.1% of the discrepancy will be reduced each year:
  - 21.1%(25.8) cents = 5.4 cents in the next month, after which an imbalance of $25.8 - 5.4 = 20.4$ cents will remain.
  - In the second month, 21.1% of the remaining 20.4 cent discrepancy, or 4.3 cents, will be eliminated; a 16.1 cent discrepancy will remain. The adjustment will continue, with smaller and smaller increments remaining, as the series reach their new equilibrium levels.

Moreover, remember that our ARIMA estimation of $Sales_t$ indicated that it included both AR(2) and MA(2) components. Perhaps not surprisingly, the ACF for the estimated residuals from Equation (10.8) indicate remaining autocorrelation, an inference that is supported by the Q-test. It appears that including a single lag of each variable is insufficient to model the autocorrelation in the $Sales_t$ series.

We will see that both of these issues – the failure to control for the reciprocal relationship and the need to include additional lags to cover the autocorrelation – are readily addressed by the Johansen method of estimating ECMs, which will be presented in the context of VAR models.

In the meantime, let's estimate the reciprocal relationship. From Equation (10.4), we would infer that in their long-run equilibrium relationship, each additional dollar of sales revenue leads, on average, to 46.2 cents in advertising spending. Saving the estimated residual $\hat{u}2_{t-1}$ from that equation and estimating the associated ECM yields the following estimates:

$$\Delta \text{Advertising}_t = \underset{(23.7)}{-6.97} + \underset{(0.094)}{0.332}\, \Delta \text{Sales}_t - \underset{(0.112)}{0.494}\, \hat{u}2_{t-1} + \hat{e}2_t \tag{10.9}$$

where standard errors of the estimates are presented in parentheses.

- In equilibrium, $u2_{t-1} = 0$. $\alpha_1 = 0.494$ indicates that a dollar increase in sales from year $t-1$ to year t is associated, on average, with a 33.2-cent increase in spending on advertising during the same period. Notice that such an increase would throw the relationship off its long-run equilibrium path; since the increase in advertising spending was more than a third less than the 46.2 cents we would expect in long-run equilibrium.

- $\hat{\alpha}_2 = -0.494$ indicates that 49.4% , or almost half of the discrepancy, will be reduced each year:
  - 49.4%(13.0) cents = 6.4 cents in the next month, after which an imbalance of $13.0 - 6.4 = 6.6$ cents will remain.
  - In the second month, 49.4% of the remaining 6.6 cent discrepancy, or 3.3 cents, will be eliminated; a 3.3 cent discrepancy will remain. The adjustment will continue, with smaller and smaller increments remaining, as the series reach their new equilibrium levels.

In this case also, we expect that our estimates of the effects of changes in sales on advertising are inflated by simultaneity bias – i.e., they are too high because we did not control for the effect of advertising on sales when estimating the effect of sales on advertising. We will resume this analysis in Chapters 12 and 13, where the VAR framework allows us to assess and control for the reciprocity between sales and advertising and properly estimate each relationship while controlling for the other.

10.4.3 Example 2: What Moves Policy Sentiment?

ECMs need not be restricted to the bivariate case. In the 1993 paper in which he introduced ECM models to the political science community, Robert Durr examined the relationship between U.S. "domestic policy sentiment" (DPS), economic expectations (EE), and federal outcomes (FedPol) from the second quarter of 1968 through the first quarter of 1988.[5] DPS is a heavily-studied measure of the relative "liberalness" of the national mood over time, as indicated by a series created by aggregating several sets of survey responses covering 60 different policy issues. (Durr 1993: 160). He hypothesized that it would exhibit cyclical behavior driven by changes in economic expectations and federal outcomes – that DPS would be more liberal, e.g., favoring expanded social welfare programs, during times when Americans were optimistic about their economic future and when federal taxes and expenditures were relatively low, but become more conservative when Americans were less confident about their economic futures and when federal taxes and expenditures were relatively high.

After verifying that all three series were I(1), he estimated a regression of DPS on economic expectations, federal outcomes and a dummy variable for

5 He also included a dummy variable for the Vietnamese War.

the Vietnamese War. In the long run, a one-point increase in the optimism indicated by the EE variable was associated with an increase of 0.39 in DPS. He was able to reject the null hypothesis that the estimated residual u_t was non-stationary, and estimated the following ECM[6]:

$$\Delta DPS_t = 40.55 + 0.16\Delta EE_t - 14.34\Delta FedPol_{t-1} - 0.67\hat{u}_{t-1} + \hat{e}_t \tag{10.10}$$

- In long-run equilibrium, $\hat{u}_{t-1} = 0$, so the short-term effect of a change in economic expectations is conveyed by the coefficient 0.16. For example, a 5-point increase in Americans' economic optimism from $t-1$ to t is estimated to be accompanied by a $(0.16)(5) = 0.8$-point liberal shift in DPS over the same period. Since this 0.8 shift is substantially lower than the long-run equilibrium relationship of a 0.39-point liberal shift for each 1-point increase in EE (or 1.95 point liberal shift associated with a 5-point increase in EE), it disrupts the equilibrium; DPS is more conservative than expected given the estimated long-run relationship.
- Approximately 2 / 3 of the adjustment back to equilibrium will occur in the first quarter following the shock; i.e., there will be a liberal shift of $0.67(1.95) = 1.31$ points in the next quarter, leaving a remaining discrepancy of 0.64. 2 / 3 of this remaining discrepancy is expected to be eliminated in the second quarter following the shock; i.e., we expect a 0.43 liberal shift. This pattern of adjustment will continue as the series reach their new equilibrium levels.

Notice the importance of the size of the adjustment parameter. In the Lydia Pinkham sales and advertising, the parameter was estimated to be –0.211; little more than 1 / 5 of the adjustment occurred in the period immediately following the shock. In the case of national policy sentiment, though, the estimated parameter was –0.67; a full 2 / 3 of the adjustment occurred in the period following the shock. Adjustment back to equilibrium will occur more quickly the larger the adjustment parameter is in absolute value.

10.4.4 Limitations of the Engle-Granger Approach

Although the Engle-Granger approach has remained popular, probably due to its simplicity, its limitations render it inferior to the Johansen method that will be covered in Chapters 12 and 13. First, as we will see when we compare the

6 It is unclear why he chose to lag $\Delta FedPol$ in the ECM equation but did not lag $FedPol_t$ in the long-run equilibrium equation.

Lydia Pinkham results from this section to the results obtained from the Johansen method, it does not control for reciprocal relationships between variables and so point estimates may be plagued by simultaneity bias. Second, while including enough lags to cover autocorrelation is a routine exercise in the VAR context, it is often overlooked when using the Engle-Granger method. Finally, because $\hat{u}_{t-1}$ in our Engle-Granger ECM equation is an estimate, it is not clear that the standard significance tests apply.

10.5 Summary

- Nonsense regressions of non-stationary series are typically statistically significant because the t-test from OLS regression is biased toward falsely rejecting the null hypothesis. Warning signs include the combination of a high R^2 and a low DW statistic.
- Solutions include the use of ADL models, the Cochrane-Orcutt method correcting for first-order autocorrelation, and differencing the series.
- An ECM, which characterizes both the long-run equilibrium between variables and the short-term impact of changes, is also a solution if the series are cointegrated.
- Series are said to be cointegrated if they are integrated of the same order and there exists a linear combination of the series that is integrated at a lower order.
- Engle and Granger provided a simple method of estimating ECMs that dovetails with cointegration tests.
- The Johansen method, computed in the context of VAR models, confers several advantages and will be introduced in Chapters 12 and 13.

References

Banerjee A., Dolado J.J., Hendry D.F., and Smith G.W. (1986). Exploring equilibrium relationships in econometrics through static models: some Monte Carlo evidence. *Oxford Bulletin of Economics and Statistics* 48: 253–277.

Durr, R.H. (1993). What Moves Policy Sentiment? *The American Political Science Review* 87,1 (Mar.): 158–170.

Engle, R.F., and Granger, C.W.J (1987). Cointegration and Error-Correction: Representation, Estimation, and Testing. *Econometrica* 55 (March): 251–276.

Engle, R.F., Granger, C.W.J. (eds) (1991). *Long Run Economic Relations: Readings in Cointegration.* Oxford: Oxford University Press.

Granger C.W.J. (1983). Co-Integrated Variables and Error-Correcting Models. *Unpublished University of California, San Diego, Discussion Paper* 83–13.

Granger C.W.J. and Newbold P. (1974). Spurious Regressions in Econometrics. *Journal of Econometrics* 2:111–120.

Hamilton J.D. (1994). Time Series Analysis. Princeton: Princeton University Press.

MacKinnon J.G. (2010). Critical Values for Cointegration Tests. *Queen's Economics Department Working Paper, No. 1227*. Queen's University, Department of Economics, Kingston (Ontario). downloaded from https://www.econstor.eu/bitstream/10419/67744/1/616664753.pdf February 16, 2021.

Murray M. (2008). A Drunk and Her Dog: An Illustration of Cointegration and Error Correction. *The American Statistician* 48,1 (Feb.): 37–39.

Phillips P.C.B. (1985). Understanding Spurious Regressions in Econometrics. Cowles Foundation Discussion Paper No. 757. New Haven: Cowles Foundation for Research in Economics.

11

The LSE Approach: Encompassing, General-to-Specific Modeling, and Forecasting Success

> *The applied econometrician is like a farmer who notices that the yield is somewhat higher under trees where birds roost, and he uses this as evidence that bird droppings increase yields. However… another farmer… objects that he used the same data but came up with the conclusion that moderate amounts of shade increase yields.*
>
> (Leamer, 1983: 31)

General-to-specific modeling is a hallmark of the LSE School approach to time series analysis, and its introduction in Chapters 8–10 likely seemed a radical departure from the ARIMA modeling covered in Chapters 4–7. In 1978, however, James Davidson, David Hendry, Frank Srba, and Stephen Yeo (DHSY) published a canonical paper that "…illustrate[d] the advantage of using a wide range of different techniques (including both 'econometric' and 'time series' methods) when analyzing aggregate data."[1] While they examined autocorrelation functions and consciously modeled the autocorrelation of their series in the tradition of ARIMA modelers, they applied the principle of "general-to-specific" modeling in the same way that it was applied in the COMFAC case and tested, rather than assumed, the implied restrictions. An important side effect of their modeling approach is that because their error term is white noise, dynamics are explicitly modeled using the included variables instead of hidden in an ARMA error term.

An equally important feature of the DHSY approach was a different type of application of the general-to-specific modeling principle. In the same way that Leamer's fabled Aviophiles and Luminists marshalled evidence from the same data to support their competing explanations for the higher agricultural yields under trees, microeconomists had proffered competing theories of the relationship between consumption and income and, in some cases, had used the same data to reach different conclusions. The first step in DHSY's analysis was to write

---

1 Davidson et al. (1978: 663).

Applied Time Series Analysis for the Social Sciences: Specification, Estimation, and Inference, First Edition. Regina M. Baker.

Companion website: www.wiley.com/go/baker

a model that encompassed the competing theories and permitted a critical test between them.

Finally, they urged examination of statistical criteria beyond R^2s, t-tests for coefficients and the Durbin–Watson statistic. Instead, they carefully examined their models' forecasting success; compared the explanatory power of restricted and unrestricted models; and assessed parameter constancy, the stability of coefficient estimates throughout the time under study.

The economic theory required for a thorough presentation of this remarkable paper is beyond the scope of this book. Charemza and Deadmans (1997) provide a simplified exposition of both the economics and DHSY's modeling process, and the presentation of competing models, below, leans heavily on their work. Readers with limited economics training who are looking for a more detailed presentation are referred to their Chapter 3. DHSY's interplay between economic theory and time series modeling is exemplary; readers who do have sufficient economics training to follow are encouraged to read the original paper in its entirety.

The first section of this chapter introduces the problem: competing models of the consumption function are supported by data analyses that meet common criteria for acceptability. Section 11.2 introduces additional criteria for consideration: forecast success and a measure of parameter non-constancy. A brief section explaining the nature of DHSY's "encompassing model" follows. Section 11.4 traces DHSY's attention to the time series properties of the data, and Section 11.5 explains their use of the dynamic properties of their penultimate candidate model to motivate further refinement.

11.1 Competing Models of the Consumption Function

The relationship between people's income and how much they spend (consume), vs. save, is a central focus of economic theory. Competing theories are commonly tested using aggregate data, and DHSY cite nine "representative" studies drawing different inferences. Following Charemza and Deadmans, we will focus on two competing theories of the relationship between consumption and income: Keynes' Absolute Income Hypothesis (AIH), and an alternative Permanent Income Hypothesis (PIH). Both are theoretically plausible, and following standard methodologies, both were supported by the same data in the sense that t-statistics on the parameters of interest indicated statistical significance at the 0.05 level, and R^2s indicated high explanatory power, and Durbin–Watson statistics indicated minimal autocorrelation. This is highly unsatisfying and the motivation for DHSY's pursuit of a better modeling strategy and better criteria for success.

11.1.1 Keynes' Absolute Income Hypothesis (AIH)

Keynes posited that consumption in each period would be a roughly fixed proportion of income in each period. Following the Cowles Commission methodology of writing a linear equation representing this hypothesis and adding a random error term, we write

$$C_t = \beta_0 + \beta_1 Y_t + \varepsilon_t,$$

where

- C_t is consumption during period t,
- Y_t is income during period t,
- $\varepsilon_t \sim N(0, \sigma^2)$, and
- the marginal propensity to consume(β_1) is less than 1.

11.1.2 The Permanent Income Hypothesis (PIH)

The PIH was motivated by a recurrent finding in the econometrics literature. In the language of economists, the average propensity to consume (rather than save) over the long run was higher than the marginal propensity to consume in the short term. In lay terms, suppose that over the long run, consumers spend ninety percent of their income. If they receive a sudden windfall one period, however, they tended to spend less than ninety percent of it immediately. The PIH incorporates the notion that impermanent increases in income, like lottery winnings or occasional gifts, are not consumed the same way that permanent income is. Instead, permanent consumption is modeled as a function of permanent income, and transitory consumption and transitory income are uncorrelated – i.e., short-term movements off the expected income path are converted into savings or expenditure on durable goods. rather than consumption.

With the application of some additional economic theory, the hypothesis can be modeled as follows:

$$C_t = \beta_0 + \beta_1 Y_t + \beta_2 C_{t-1} + v_t,$$

where C_t and Y_t are defined as above, and $v_t \sim N(0, \sigma^2)$.

The results of OLS estimations of both models are reported in Table 11.1. Although available, data from 1971 to 1975 were not included in the estimations so that they could be compared with forecasts. In order to facilitate comparisons with competing models that include seasonal and regular differencing, the first five observations were also dropped, so the included cases are quarterly data from 1958Q2 through 1970Q4.

Both models pass standard criteria. R^2s exceeding 0.9 for both indicate high explanatory power. p-values for the coefficients of interest are well below 0.05,

Table 11.1 Results for OLS estimations of the AIH and PIH models.

Model	β_0	β_1	β_2	R^2	DW
AIH	967	0.73		0.91	2.17
	(< 0.001)	(< 0.001)			
PIH	1250	0.86	−0.20	0.92	1.79
	(< 0.001)	(< 0.001)	(0.018)		

P-values are reported in parentheses.
DW is the Durbin–Watson statistic.
$n = 51$.
Source: Data source: Charemza and Deadmans 1997 / Edward Elgar.

and Durbin–Watson statistics near 2 suggest there's little autocorrelation in the residuals. The estimated negative coefficient on the lagged consumption variable (C_{t-1}) in the PIH model, counter to the theoretical prediction, is the only immediate sign of trouble.

11.2 Additional Criteria: Forecast Success and Parameter Constancy

Fortunately, both models fail by other criteria, however. Out-of-sample forecasts are poor in both cases, and neither model passes tests for parameter constancy.

Recall that the time period 1971–1975 was excluded from the estimations of the AIH and PIH models so that data from this period could be used to examine forecasts. 95% Forecast intervals for Consumption generated from the AIH and PIH estimations are compared to the realized Consumption values from 1971 through 1975 in Figure 11.1. In Figure 11.1a, we see that a substantial number of the realized values are outside the 95% forecast interval. Indeed, careful examination reveals that three of the five first-quarter observations are well below the interval, suggesting that we may have failed to model a seasonal component. Forecasts from the PIH model follow a similar pattern.

One need not rely entirely on visual comparison of forecast values to realize consumption values, though, to infer that the estimated model does not perform well in the post-estimation period. Two tests for parameter non-constancy follow. The first is a version of the Chow test that was first seen in Chapter 9. The second is the Forecast χ^2 test. Both test the null hypothesis that all parameter values are the same for the sample and forecast periods.

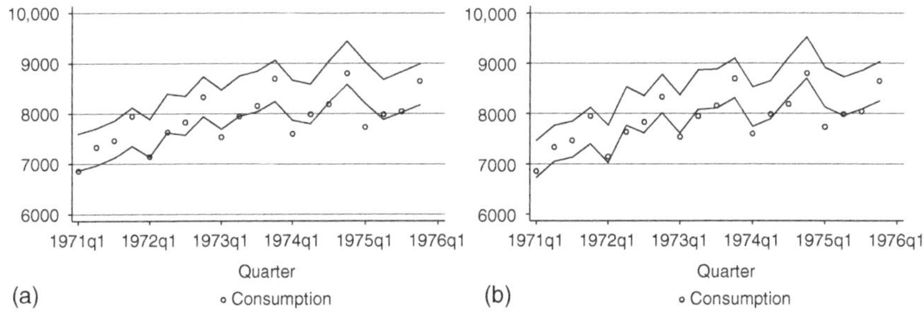

Figure 11.1 Forecast intervals and realized consumption values for AIH and PIH models. (a) 95% forecast interval and actual consumption for the AIH model. (b) 95% forecast interval and actual consumption for the PIH model..

11.2.1 A Chow Test for Parameter Non-constancy

The Chow test for parameter non-constancy is calculated as

$$\frac{(ESS^* - ESS)/(T^* - T)}{ESS/(T-k)} \sim F_{T^*-T,T-k},$$

where

- ESS and are the error (residual) sum of squares from the model using data only through period T,
- ESS^* is the error (residual) sum of squares from the model using all data, including the forecast period,
- T is the last point in the "in-sample" portion of the series,
- T^* is the end of the entire series (includes "in-sample" and forecast period), and
- k is the number of coefficients estimated, including the constant term.

The null hypothesis is that the

The Error Sums of Squares for the AIH and PIH estimations are presented in Table 11.2.

In the case of the AIH estimation, then, the test statistic is computed as

$$(3{,}256{,}656 - 1{,}651{,}709)/201{,}651{,}709/(51-2) = 2.38.$$

Table 11.2 Error (residual) sums of squares for the AIH and PIH estimations.

Model	*ESS*	*ESS**
AIH	1,651,709	3,256,656
PIH	1,469,171	2,778,001

If the test value exceeds the cutpoint for the chosen level of statistical significance, we are able to reject the null hypothesis that parameters are the same for the in-sample and post-estimation periods even at a stringent.01 level of statistical significance.

11.2.2 The Forecast χ^2 Test for Parameter Non-constancy

A second test of parameter non-constancy makes direct use of the forecast errors. The test statistics for the Forecast χ^2 test is

$$\frac{\sum_{t=T+1}^{T^*} fe_t^2}{\sigma^2},$$

where

- $T+1$ is the first point in the forecast portion of the series,
- T^* is the last period of the forecast,

Table 11.3 Realized and forecast values from AIH and PIH estimations.

Quarter	Consumption	AIH Forecast	PIH Forecast
1971q1	6855	7234	7100
1971q2	7335	7339	7410
1971q3	7467	7491	7494
1971q4	7952	7740	7761
1972q1	7147	7514	7397
1972q2	7636	8012	8146
1972q3	7829	7961	7988
1972q4	8332	8337	8394
1973q1	7539	8082	7992
1973q2	7948	8355	8473
1973q3	8157	8440	8492
1973q4	8691	8654	8702
1974q1	7601	8265	8137
1974q2	7985	8193	8269
1974q3	8186	8634	8713
1974q4	8798	9006	9112
1975q1	7735	8609	8522
1975q2	7984	8280	8345
1975q3	8045	8428	8470
1975q4	8646	8581	8639

- fe_t is the forecast error, the difference between the forecast value and the actual value, and
- σ is the standard error of the regression from the original estimation.

The actual and forecast values for consumption from the AIH and PIH estimations are presented in Table 11.3.

Forecast errors are calculated as $Consumption_t - Forecast_t$ for each quarter. Each of these is squared, and the sum of these squares becomes the numerator for the Forecast χ^2 test statistic. For the AIH estimation, the standard error of the regression (σ) was 183.6. Then the test statistic becomes

$$\frac{2,755,287}{183.6^2} = 81.7.$$

Comparing this to the χ^2 distribution with 20 degrees of freedom yields $p = 0.001$. Again, we reject the null hypothesis of parameter constancy across periods even at the 99

11.3 A Sketch of the DHSY Process

DHSY continue their exposition by sketching a process for model development that is characterized by three critical features: encompassing tests, explicit modeling of time series properties, and general-to-specific modeling. Each of these is explained in turn.

11.3.1 Encompassing and Critical Tests

A critical aspect of DHSY's process was its application of general-to-specific modeling to problems such as that posed by Leamer in his fable about the competing Aviophile and Luminist explanations of increased agricultural yields under trees.

DHSY advise writing and testing a more general statistical model that encompasses both specific hypotheses and provides a critical test between them. While the concept of encompassing has been further developed, the simplest form is a model in which competing theories are nested within the estimating equation. Notice that in this case, the AIH hypothesis is nested within the PIH model: the AIH is a specific case of the estimating equation where $\hat{\beta}_2 = 0$, leading to a simple critical test: if the AIH is correct, $\hat{\beta}_2$ should not be statistically significant. The result in this case is unsatisfying. The coefficient estimate is statistically significant, which might favor the PIH over the AIH; however, the sign is negative, inconsistent with the PIH model.

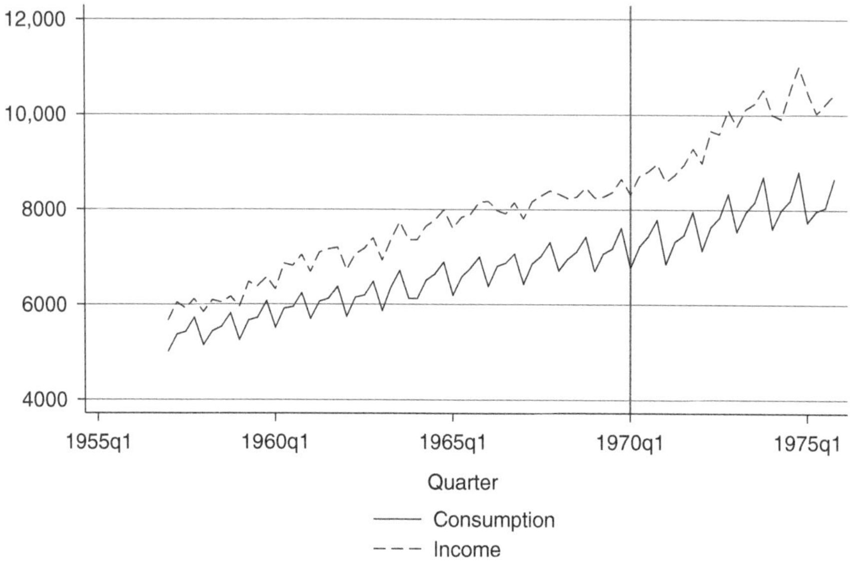

Figure 11.2 U.K. consumption and income, 1957–1975.

11.3.2 Time Series Properties and General-to-Specific Modeling

Our analysis thus far is a cautionary tale illustrating the importance of modeling the time series properties of our series.

A plot of the relationship between consumption and income is presented in Figure 11.2.

Some properties that we've examined in earlier chapters should be immediately apparent:

- both series are mean non-stationary,
- both series exhibit strong seasonal patterns, and
- the variance of both series appears to increase with the mean of the series.

After logging both series to address the variance non-stationarity and checking different combinations of differencing and seasonal differencing, DHSY arrive at this intermediate specification:

$$\Delta_4 lnC_t = \delta_0 + (\delta_1 + \delta_2)\Delta_4 lnY_t - \delta_2 \Delta_1 \Delta_4 lnY_t + e_t, \tag{11.1}$$

where

- the Δ_4 operator is the seasonal quarterly difference operator $1 - L_4$, (not a 4th difference operator $(1-L)^4$),

- δ_0, δ_1, and δ_2 are parameters to be estimated, and
- $e_t \sim N(0,\sigma^2)$.[2]

Before accepting this specification, however, they expand the difference operators in Equation (11.1) to rewrite the equation in its general form[3]:

$$lnC_t = \gamma_1 lnY_t + \gamma_2 lnY_{t-4} - \gamma_3 lnY_{t-1} + \gamma 4 lnY_{t-5} + \gamma_5 lnC_{t-4} + \gamma_6 + e_t, \tag{11.2}$$

and note the implied restrictions:

- $\gamma_1 = -\gamma_2$,
- $\gamma_3 = -\gamma_4$, and
- $\gamma_5 = 1$.

The restrictions are tested using the following variant of the Chow test:

$$\frac{(ESS_R - ESS_U)/K_r}{ESS_U/(T-K)} \sim F_{K_r,T-K},$$

where

- T is the sample size,
- ESS_R is the error (residual) sum of squares from the restricted model (e.g., Equation (11.1)),
- RSS_U is the error (residual) sum of squares from the unrestricted model (e.g., Equation (11.2)),
- K_r is the number of restrictions, and
- K is the number of parameters (coefficients) estimated in the unrestricted regression. (Include the intercept in K.)

11.3.3 Incorporation of the Adjustment (Error-Correction) Mechanism

Part of their evaluation of candidate models involves checking their residuals to confirm that they're white noise. As a consequence, they are able to trace the dynamics implied by their models. They find their penultimate model suggests non-credible behavior – it does not result in a stable equilibrium – and introduce an error correction piece to address this. Their final model has a white-noise residual, credible dynamic properties, and high forecasting success; and passes tests of parameter non-constancy.

---

2 DHSY 1978, Equation (19), p. 673.

3 DHSY 1978, Equation (20), p. 674.

11.4 Summary

- The common approach of writing an equation that represents a theoretical process, estimating a linear regression, and then examining the associated t-tests, R^2, and Durbin–Watson statistic can yield results that are both credible and support different conclusions.
- Additional criteria for time series models include tests for parameter non-constancy and the performance of out-of-sample forecasts.
- Writing an encompassing model permits a critical test to discern between competing hypotheses.
- Modeling the time series properties of the series is important for mitigating spurious results due to excluded (sometimes unmeasurable) variables.
- Better models can be achieved by modeling the time series properties of the series but testing the implied restrictions before accepting them.
- Evaluating the theoretical implications of candidate models along with estimation results is critical to developing a sound model.

References

Charemza W.W. and Deadman D.F. (1997). *New Directions in Econometric Practice: General to Specific Modelling, Cointegration and Vector Autoregression,* 2nd ed. Cheltenham UK: Edward Elgar.

Davidson J.E., Hendry D.F., Srba F., and Yeo S. (1978). Econometric Modelling of the Aggregate Time-Series Relationship Between Consumers' Expenditure and Income in the United Kingdom. *The Economic Journal* 88(352):661–692.

Leamer E.E. (1983). Let's Take the Con out of Econometrics. *The American Economic Review* 73(1):31–43.

12

A Brief Introduction to Vector Autoregression

To the uninitiated, Vector Autoregression (VAR) often appears appallingly atheoretical; the approach is quite a departure from the recommendation to write a statistical model based on theory and then estimate the associated parameters. To use VAR to assess the relationship between consumption and income, for example, one would begin with the following model:

$$C_t = \alpha_1 + \beta_{11}Y_{t-1} + \beta_{12}Y_{t-2} + \cdots \beta_{1k}Y_{t-k} + \gamma_{11}C_{t-1} + \gamma_{12}C_{t-2} + \cdots + \gamma_{1k}C_{t-k} + u_1 \quad (12.1)$$

and

$$Y_t = \alpha_2 + \beta_{21}C_{t-1} + \beta_{22}C_{t-2} + \cdots \beta_{2k}C_{t-k} + \gamma_{21}Y_{t-1} + \gamma_{22}Y_{t-2} + \cdots + \gamma_{2k}Y_{t-k} + u_2, \quad (12.2)$$

where

- C_t is consumption,
- Y_t is income,
- α_1 and α_2 are constant terms to be estimated,
- the γ_{1k}s and γ_{2k}s are coefficients to be estimated, and
- $u_1 \sim \mathrm{N}(0,\sigma_1^2)$ and $u_2 \sim \mathrm{N}(0,\sigma_2^2)$.

As mentioned in Chapter 1, however, forecasts from VAR models have typically outperformed those of "theory-driven" dynamic simultaneous models. By now, you might immediately recognize that the system of equations represented by Equations (12.1) and (12.2) are a generalization of linear models of C_t and Y_t. The VAR framework allows us to test restrictions rather than assume them. Indeed, the term "theory-laden" is the retort associated with "The Sims Critique":

Applied Time Series Analysis for the Social Sciences: Specification, Estimation, and Inference, First Edition. Regina M. Baker.

Companion website: www.wiley.com/go/baker

Christopher Sims' concern that the assumptions needed to achieve identification in many macroeconomic models are questionable and compromising.[1]

VAR was originally introduced by Sims to address common problems in the estimation of simultaneous equations models involving time series data (dynamic simultaneous equations models) (Sims, 1980). The key feature of these models is that at least one variable that appears as an explanatory variable in one equation is also "endogenously" determined in the system – it appears as a dependent variable in another equation. Recall that for an equation in a multiple-equation system to be estimated, it must be "identified" – that is, the order and rank conditions concerning the other variables in the model must be met. Unfortunately, theory does not always lead to systems of identified equations, and faced with the choice between tinkering (atheoretically) with the model to achieve identification and dropping projects entirely, many researchers tend to do the former. It would be naive to suggest that theoretical revisionism in the face of uncooperative data is rare practice, and its effects can be particularly egregious in the case of time series data. In the case of cross-sectional studies, the direction of the bias plaguing the coefficients of variables remaining in a misspecified equation is relatively simple to predict. Dynamic simultaneous equations models, though, are notoriously sensitive to model specification; dropping one variable from an equation frequently leads to sign changes in the coefficients associated with remaining variables.

VAR mitigates these problems because it does not require analysts to decide *a priori* either which variables are endogenous and which are exogenous, or which variables should appear in which equation. It allows for the possibility that all variables are determined endogenously, and all variables (and lagged versions of them) are candidates for included independent variables. It may appear at first that a disadvantage is the inability to interpret coefficient estimates as marginal effects, but as emphasized in Chapter 9, coefficient estimates from many commonly used dynamic models cannot be interpreted as marginal effects. VAR permits the estimation of Impulse Response Functions (IRFs), though, which trace the effects of a change in each explanatory variable on changes in each dependent variable over time. VAR also provides the foundation for Granger Causality tests, tests for a narrow definition of "causality" as the extent to which the dependent variable in one equation improves forecasts of a dependent variable in another equation.

This chapter provides only a brief introduction to the VAR framework. Section 12.1 provides a more detailed explanation of the logic, including the assumptions required to estimate multiple equation models. Section 12.3 introduces the estimation framework. Granger Causality tests are presented in Section 12.4, and impulse response functions are introduced in Section 12.5. Lydia Pinkham's sales and advertising data are used for illustration.

---

1 Sims (1980).

12.1 VAR: Logic and Motivation

12.1.1 General-to-Specific Modeling

General-to-specific modeling has been a recurring theme through the second half of this text. Recall, for example, that DHSY estimated their generalized short-term effects model of consumption spending as

$$lnC_t = \gamma_1 lnY_t + \gamma_2 lnY_{t-4} - \gamma_3 lnY_{t-1} + \gamma 4 lnY_{t-5} + \gamma_5 lnC_{t-4} + \gamma_6 + e_t. \quad (12.3)$$

Notice that the model includes terms such as lnY_{t-4} and lnY_{t-5}. The VAR setup can be thought of as a more general form, where the omissions of intermediate lags of lnY_t such as lnY_{t-2} and lnY_{t-3} are "zero restrictions" – incorporating the assumption that the coefficients on these lags are 0 without testing that assumption.

A more serious concern is the failure to consider the reciprocal path – the effect of consumption spending on income (through the demand function).

12.1.2 Identification in Multiple Equation Models

But consider what we need to do in order to estimate a standard model in which both Consumption and Income are treated as endogenously determined. Let us begin with the following system:

$$C_t = \alpha_0 + \alpha_1 Y_t + e_{1t}.$$
$$Y_t = \beta_0 + \beta_1 C_t + \beta_2 I_t + \beta_3 G_t + e_{2t}.$$[2]

Estimation involves estimating a reduced form:

$$C_t = \frac{\alpha_0 + \alpha_1\beta_0}{1-\alpha_1\beta_1} + \frac{\alpha_1\beta_2}{1-\alpha_1\beta_1} I_t + \frac{\alpha_1\beta_3}{1-\alpha_1\beta_1} G_t + \frac{e_{1t} + \alpha_1 e_{2t}}{1-\alpha_1\beta_1}.$$

$$Y_t = \frac{\beta_0 + \alpha_0\beta_1}{1-\alpha_1\beta_1} + \frac{\beta_2}{1-\alpha_1\beta_1} I_t + \frac{\beta_3}{1-\alpha_1\beta_1} G_t + \frac{\beta_1 e_{1t} + e_{2t}}{1-\alpha_1\beta_1}.$$

The "identification problem" refers to our ability to solve for the α and β coefficients after estimating the reduced form model.

- We can retrieve estimates of the structural parameters (α's and β's) from our estimation of the reduced form equations only if we meet the *order condition* and the *rank* condition.
- Sometimes equations will be left unidentified.
- Worse, sometimes researchers will "adjust" their theory to achieve identification.

2 Apologies to anyone offended by the notion of estimating an accounting identity.

If we are able to solve the equations to get unique solutions for the original coefficients (α's and β's) we say that the equation is "identified."

The order condition requires that the number of exogenous variables excluded from a particular equation must equal at least the number of endogenous variables included as explanatory variables in the equation.

To understand how this invites ad hoc adjustments to theory, consider an under-identified equation.

Suppose we begin with the following system.

$$Y_{1i} = \beta_{10} + \gamma_{12}Y_{2i} + \beta_{11}X_{1i} + e_{1i}.$$
$$Y_{2i} = \beta_{20} + \gamma_{21}Y_{1i} + e_{2i}.$$

We can obtain the reduced form equation for Y_{1i} by substituting the second equation into the first:

$$\begin{aligned} Y_{1i} &= \beta_{10} + \gamma_{12}(\beta_{20} + \gamma_{21}Y_{1i} + e_{2i}) + \beta_{11}X_{1i} + e_{1i} \\ &= \beta_{10} + \gamma_{12}\beta_{20} + \gamma_{12}\gamma_{21}Y_{1i} + \gamma_{12}e_{2i} + \beta_{11}X_{1i} + e_{1i} \\ &= \frac{\beta_{10} + \gamma_{12}\beta_{20}}{1-\gamma_{12}\gamma_{21}} + \frac{\beta_{11}}{1-\gamma_{12}\gamma_{21}}X_{1i} + \frac{\gamma_{12}e_{2i} + e_{1i}}{1-\gamma_{12}\gamma_{21}} \\ &= \pi_{10} + \pi_{11}X_{1i} + u_{1i}, \end{aligned}$$

where

$$\pi_{10} = \frac{\beta_{10} + \gamma_{12}\beta_{20}}{1-\gamma_{12}\gamma_{21}}$$

$$\pi_{11} = \frac{\beta_{11}}{1-\gamma_{12}\gamma_{21}}, \text{ and}$$

$$u_{1i} = \frac{\gamma_{12}e_{2i} + e_{1i}}{1-\gamma_{12}\gamma_{21}}.$$

Similarly, we can obtain the reduced form equation for Y_{2i} by substituting the first equation in the system into the second:

$$\begin{aligned} Y_{2i} &= \beta_{20} + \gamma_{21}(\beta_{10} + \gamma_{12}Y_{2i} + \beta_{11}X_{1i} + e_{1i}) + e_{2i} \\ &= \beta_{20} + \gamma_{21}\beta_{10} + \gamma_{21}\gamma_{12}Y_{2i} + \gamma_{21}\beta_{11}X_{1i} + \gamma_{21}e_{1i} + e_{2i} \\ &= \frac{\beta_{20} + \gamma_{21}\beta_{10}}{1-\gamma_{12}\gamma_{21}} + \frac{\gamma_{21}\beta_{11}}{1-\gamma_{12}\gamma_{21}}X_{1i} + \frac{\gamma_{21}e_{1i} + e_{2i}}{1-\gamma_{12}\gamma_{21}}, \end{aligned}$$

or

$$Y_{2i} = \pi_{20} + \pi_{21}X_{1i} + u_{20},$$

where

$$\pi_{20} = \frac{\beta_{20} + \gamma_{21}\beta_{10}}{1-\gamma_{12}\gamma_{21}}$$

$$\pi_{21} = \frac{\gamma_{21}\beta_{11}}{1-\gamma_{12}\gamma_{21}}, \text{ and}$$
$$u_{2i} = \frac{\gamma_{21}e_{1i} + e_{2i}}{1-\gamma_{12}\gamma_{21}}.$$

Notice what has happened. If we run OLS, we will get 4 coefficient estimates (the 4 π's). However, we have 5 unknowns – β_{10}, γ_{12}, β_{11}, β_{20}, and γ_{21}.This means that we can't solve for all of the original parameters. In this system of equations, the Y_{2i} equation passes the order condition, but the Y_{1i} equation does not.

To meet the order condition, one more exogenous variable must be added. Moreover, the order condition is a necessary, but not sufficient, condition for identification. The rank condition is a necessary and sufficient condition, but is even more stringent. Ad hoc tinkering with the original specification in order to meet the identification criteria needed for estimation can compromise the model in unpredictable ways.

12.2 Estimating VAR Models

VAR models can be estimated without these ad hoc assumptions.

12.2.1 Basic Setup: From the Structural VAR to the Reduced Form

Suppose we begin with the following general structural model:

$$\begin{aligned} X_t &= c_1 + \omega_{11}X_{t-1} + \omega_{12}X_{t-2} + \omega_{13}X_{t-3} + \cdots + \omega_{1k}X_{t-k} \\ &\quad + \delta_{10}Y_t + \delta_{11}Y_{t-1} + \delta_{12}Y_{t-2} + \delta_{13}Y_{t-3} + \cdots \delta_{1k}Y_{t-k} + u_{1t}. \\ Y_t &= c_2 + \omega_{20}X_t + \omega_{21}X_{t-1} + \omega_{22}X_{t-2} + \omega_{23}X_{t-3} + \cdots + \omega_{2k}X_{t-k} \\ &\quad + \delta_{21}Y_{t-1} + \delta_{22}Y_{t-2} + \delta_{23}Y_{t-3} + \cdots \delta_{2k}Y_{t-k} + u_{2t} \end{aligned}$$

where

- X_t and Y_t are the time series of interest,
- c_1 and c_2 are constant terms to be estimated,
- the ω_{1k}s and δ_{2k}s are coefficients to be estimated, and
- $u_1 \sim (0, \sigma_1^2)$ and $u_2 \sim (0, \sigma_2^2)$.

By subtracting $\delta_{10}Y_t$ and $\omega_{20}X_t$ from both sides of the first and second equations, respectively, we can write

$$\begin{aligned} X_t - \delta_{10}Y_t &= c_1 + \omega_{11}X_{t-1} + \omega_{12}X_{t-2} + \omega_{13}X_{t-3} + \cdots + \omega_{1k}X_{t-k} \\ &\quad + \delta_{11}Y_{t-1} + \delta_{12}Y_{t-2} + \delta_{13}Y_{t-3} + \cdots \delta_{1k}Y_{t-k} + u_{1t}. \end{aligned}$$

$$Y_t - \omega_{20}X_t = c_2 + \omega_{21}X_{t-1} + \omega_{22}X_{t-2} + \omega_{23}X_{t-3} + \cdots + \omega_{2k}X_{t-k}$$
$$+ \delta_{21}Y_{t-1} + \delta_{22}Y_{t-2} + \delta_{23}Y_{t-3} + \cdots \delta_{2k}Y_{t-k} + u_{2t}.$$

In matrix form, then, the left-hand side of this system of equations can be written

$$\begin{bmatrix} 1 & -\delta_{10} \\ -\omega_{20} & 1 \end{bmatrix} \begin{bmatrix} X_t \\ Y_t \end{bmatrix}$$

To get the reduced form version of this system, multiply both sides by the inverse of

$$\begin{bmatrix} 1 & -\delta_{10} \\ -\omega_{20} & 1 \end{bmatrix}$$

This will result in a system that takes this form:

$$X_t = \alpha_1 + \beta_{11}X_{t-1} + \beta_{12}X_{t-2} + \beta_{13}X_{t-3} + \cdots + \beta_{1k}X_{t-k}$$
$$+ \gamma_{11}Y_{t-1} + \gamma_{12}Y_{t-2} + \gamma_{13}Y_{t-3} + \cdots \gamma_{1k}Y_{t-k} + e_{1t}.$$
$$Y_t = \alpha_2 + \beta_{21}X_{t-1} + \beta_{22}X_{t-2} + \beta_{23}X_{t-3} + \cdots + \beta_{2k}X_{t-k}$$
$$+ \gamma_{21}Y_{t-1} + \gamma_{22}Y_{t-2} + \gamma_{23}Y_{t-3} + \cdots \gamma_{2k}Y_{t-k} + e_{2t}.$$

- Notice that both variables (all variables, in a multivariate system) are treated equally.
- Estimation is typically done with OLS, based on the assumption that enough lags are included so that autocorrelation is not an issue. (SEEMINGLY-unrelated regression [SUR], while equally appropriate, will not add to the efficiency because all equations have the same right-hand-side [RHS] variables.)

To undertand why including enough lags will leave us with equations free of autocorrelation, remember from Wold's Theorem in Section 4.2 that the difference between any weakly stationary series and its mean can be modeled as a linear combination of random shocks, and that any moving average process can be written as an autoregressive process – i.e., an equation in lags of Y.

Then inclusion of enough lagged Y_t terms will leave you with a white-noise error term.

12.2.2 Choosing Lag Length

Since all variables in the system will appear in all equations, and the same lag length will be applied to all of them, there's tension between specifying too few lags to eliminate autocorrelation, and specifying too many to estimate anything precisely. Remember, though, that these are dynamic models. The coefficient estimates are not appropriately interpreted as marginal effects. Instead, the estimates are typically used in any of three ways: forecasting, performing Granger Causality tests, and generating impulse response functions. In all these applications, the specification bias caused by leaving autocorrelation in the error term is typically a worse problem than efficiency lost by including more variables than needed; i.e., including too many lags is preferable to including too few.

Guidelines for deciding how many lags to include are based on two types of considerations. First, the lags should cover any "season" in the data. At first blush, this would seem to imply a minimum of four lags for quarterly data and twelve lags in monthly data. In practice, however, autocorrelation is often seen in the lags immediately following the end of a season, so checking at least six lags for quarterly data and at least 15 for monthly data is advisable.

The second consideration is the explanatory power added by including additional lags. This is commonly assessed by either likelihood ratio tests or information criteria such as the AIC and SBC that were presented in Chapter 5. Do not use t-tests to decide how many lags to keep. Because lagged variables tend to be highly correlated, coefficients are typically estimated imprecisely and relying on t-tests may result in including too few lags.

12.2.3 Example: Lydia Pinkham's Sales and Advertising Data

In Chapter 7, we began by assuming that sales was our dependent variable and advertising was the independent variable. Our CCF indicated, however, that sales levels were affecting advertising. Recall that we logged both series to address variance nonstationarity. To examine the possibility of a reciprocal relationship, we begin with the following VAR model:

$$\begin{aligned} lSales_t &= \alpha_1 + \beta_{11} lSales_{t-1} + \beta_{12} lSales_{t-2} + \cdots \beta_{1k} lSales_{t-k} + \gamma_{11} lAdv_{t-1} \\ &\quad + \gamma_{12} lAdv_{t-2} + \cdots + \gamma_{1k} lAdv_{t-k} + u_1 \end{aligned} \tag{12.4}$$

and

$$\begin{aligned} lAdv_t &= \alpha_2 + \beta_{21} lAdv_{t-1} + \beta_{22} lAdv_{t-2} + \cdots \beta_{2k} lAdv_{t-k} + \gamma_{21} lSales_{t-1} \\ &\quad + \gamma_{22} lSales_{t-2} + \cdots + \gamma_{2k} lSales_{t-k} + u_2, \end{aligned} \tag{12.5}$$

where

- $lSales_t$ is the logged Sales variable at time t,
- $lAdv_t$ is the logged Advertising variable at time t,
- $u_1 \sim N(0, \sigma_1^2)$, and
- $u_2 \sim N(0, \sigma_2^2)$.

Table 12.1 AIC and SBC values for ln(Sales) and ln(Advertising).

Lag	AIC	SBC
1	−1.90224	−1.66373
2	−2.0123	−1.61477
3	−2.03156	−1.47501
4	−2.06038	−1.34483
5	−1.99693	−1.12236
6	−1.86538	−0.831805
7	−1.79982	−0.607226
8	−1.88191	−0.530309

The first decision we must make is how many lags of each variable to include in the model. Because these are annual data, we need not worry about covering a season.

The AIC and SBC for models with between one and eight lags included are presented in Table 12.1.

Recall that lower AICs and SBCs are preferred. In this case, the lowest AIC is −2.06038, observed for the model with four lags of each variable included. The lowest SBC is −1.66373, observed for the model with only one lag of each variable included. Since the SBC penalizes extra parameters more heavily than the AIC does, it's frequently the case that two criteria will suggest different parameterizations. Since specification bias is a worse problem in this application than inefficiency is, we will use four lags in our VAR model.

Each equation can now be estimated separately using OLS. After estimating the model, the residuals must be checked for autocorrelation.

12.2.4 Granger Causality

An advantage of the VAR methodology is it permits explicit tests of the order of causality.

Refer again to the simple VAR model presented in Section 12.2.

We infer that X_t does **not** Granger-cause Y_t if and only if all of the coefficients on X_{t-1} through X_{t-p} are equal to 0. The definition is sometimes presented in terms of forecasting performance. The logic is that if including lags of X_t does not improve forecasts of Y_t, X_t does not Granger-cause Y_t.

In the case of the Lydia Pinkham data, we were left wondering if advertising affects sales, and/or if sales affects advertising. The results of Wald

Table 12.2 Granger causality tests for sales and advertising.

Potential exogenous variable	Excluded variables	χ^2	Degrees of freedom	P-value
ln(Sales)	ln(Advertising)	8.1908	4	0.085
ln(Advertising)	ln(Sales)	21.317	4	< 0.0001

tests for Granger Causality from the VAR model using 4 lags are presented in Table 12.2.

In this case, we reject the null hypothesis that Sales does not affect Advertising at the $p < 0.0001$ level of significance. Since our p-value for the test of the hypothesis that Advertising does not affect Sales is 0.085, we cannot reject the null hypothesis that Advertising does not affect Sales at the 0.05 level of statistical significance.

12.2.5 Impulse Response Functions

The VAR model also affords a means of exploring dynamics. The coefficient estimates can be used to generate impulse response functions, representations of the response over time of one variable to changes in the other.

A graphical representation of the impulse response function showing the impact of a change in Sales on Advertising is presented in Figure 12.1. The graph

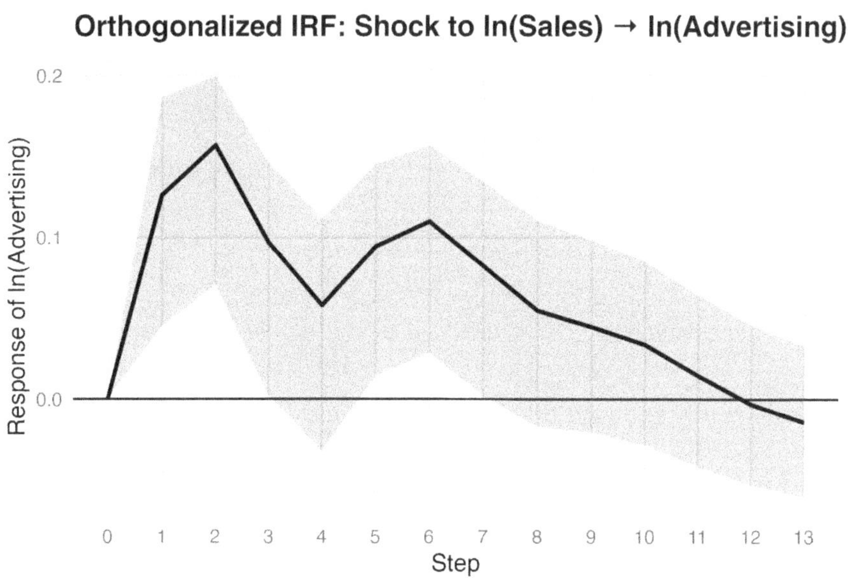

Figure 12.1 An impulse response function for the effect of changes in sales on advertising.

indicates that there is an increase in Advertising in the contemporaneous period, and a larger one in the period following the change in Sales. The effect declines afterward.

12.3 Summary

- VAR models are explicitly used to examine simultaneity, the possibility that variables that appear as explanatory variables in one equation are also dependent (endogenous) variables in a system of equations.
- They are motivated by concerns about ad hoc modeling decisions. By including all lags of all variables in the system, "zero restrictions" are eliminated.
- The number of lags to include in the VAR model is a critical modeling decision, and should be guided by the nature of any seasonality and by criteria such as the AIC and SBC that evaluate the additional information provided by the inclusion of each additional lag.
- VAR equations can be estimated using OLS; residuals should be checked for autocorrelation.
- Granger Causality tests provide a means of testing whether a variable has explanatory power on another.
- VAR estimates can be used to generate impulse response functions indicating the effect of a change in one variable on another over time.

Reference

Sims C.A. (1980). Macroeconomics and Reality. *Econometrica* 88(352):661–692.

Index

Applied Time Series Analysis for the Social Sciences: Specification, Estimation, and Inference, First Edition. Regina M. Baker.

Companion website: www.wiley.com/go/baker